Teaching Design
and Technology

DEVELOPING SCIENCE AND TECHNOLOGY EDUCATION

Series Editor: Brian Woolnough
Department of Educational Studies, University of Oxford

Current titles:

John Eggleston: *Teaching Design and Technology*
Clive Sutton: *Words, Science and Learning*

Titles in preparation include:

David Layton: *Technology's Challenge to Science Education*
Michael Poole: *Beliefs and Values in Science and Technology Education*
Keith Postlethwaite: *Teaching Science to Pupils with Special Educational Needs*
Michael Reiss: *Science Education for a Pluralist Society*
Jon Scaife and Jerry Wellington: *Information Technology in Science and Technology Education*
Joan Solomon: *Teaching Science, Technology and Society*

Teaching Design and Technology

JOHN EGGLESTON

Open University Press
Buckingham · Philadelphia

Open University Press
Celtic Court
22 Ballmoor
Buckingham
MK18 1XW

and

1900 Frost Road, Suite 101
Bristol, PA 19007, USA

First Published 1992

*A catalogue record of this book is available from
the British Library*

*Library of Congress Cataloging-in-Publication
Data*

Eggleston, S. J. (Samuel John), 1926–
    Teaching design and technology/
    John Eggleston.
        p. cm. – (Developing science and
    technology education)
    Includes index.
    ISBN 0–335–09874–6
    ISBN 0–335–09869–X (pbk.)
    1. Engineering design – Study and teaching.
  I. Title.   II. Series.
  TA174.E34   1992
  620′.0042′072 – dc20)   91–45906
                                            CIP

Typeset by Type Study, Scarborough
Printed in Great Britain by
St Edmundsbury Press, Bury St Edmunds,
Suffolk

Contents

Series editor's preface

It may seem surprising that after three decades of curriculum innovation, and with the increasing provision of centralised National Curriculum, that it is felt necessary to produce a series of books which encourage teachers and curriculum developers to continue to rethink how science and technology should be taught in schools. But teaching can never be merely the 'delivery' of someone else's 'given' curriculum. It is essentially a personal and professional business in which lively, thinking, enthusiastic teachers continue to analyse their own activities and mediate the curriculum framework to their students. If teachers ever cease to be critical of what they are doing then their teaching, and their students' learning, will become sterile.

There are still important questions which need to be addressed, questions which remain fundamental but the answers to which may vary according to the social conditions and educational priorities at a particular time.

What is the justification for teaching science and technology in our schools? For educational or vocational reasons? Providing science and technology for all, for future educated citizens, or to provide adequately prepared and motivated students to fulfil the industrial needs of the country? Will the same type of curriculum satisfactorily meet both needs or do we need a differentiated curriculum? In the past it has too readily been assumed that one type of science will meet all needs.

What should be the nature of science and technology in schools? It will need to develop both the methods and the content of the subject, the way a scientist or engineer works and the appropriate knowledge and understanding, but what is the relationship between the two? How does the student's explicit knowledge relate to investigational skill, how important is the student's tacit knowledge? In the past the holistic nature of scientific activity and the importance of affective factors such as commitment and enjoyment have been seriously undervalued in relation to the student's success.

And, of particular concern to this series, what is the relationship between science and technology? In some countries the scientific nature of technology and the technological aspects of science make the subjects a natural continuum. In others the curriculum structures have separated the two leaving the teachers to develop appropriate links. Underlying this series is the belief that science and technology have an important interdependence and thus many of the books will be appropriate to teachers of both science and technology.

Professor John Eggleston has been a central influence in the development of technology as a distinct subject over the last two decades. His analysis of its genesis and its current state in schools provides important insights and indicators for its future developments.

We hope that this book, and the series as a whole, will help many teachers to develop their science and technological education in ways that are both satisfying to themselves and stimulating to their students.

Brian E. Woolnough

Preface

In this book I have tried to relate the momentous events in Design and Technology education in recent years to the fundamental issues that underlie them. The challenge to do this in a way that will be accessible and immediately helpful is welcome – you, the reader, will judge how successful I have been. I am grateful to the publishers and the series editor, Brian Woolnough for the stimulus to undertake this task. They have been helpful throughout as have all other colleagues at the University of Warwick. Officers of the National Curriculum Council and the Schools Examinations and Assessment Council have discussed issues helpfully. Margaret Handy has kindly prepared the manuscript through numerous drafts. I am grateful to them all – but I accept total responsibility for all the faults of the final product.

John Eggleston

CHAPTER 1

The coming of Design and Technology

This chapter reviews the provenance of Design and Technology in schools and traces its growth from traditional roots through to the complex and dramatic chain of recent events which have brought recognition and high status.

Design and Technology is one of the fastest growth areas of the contemporary school curriculum. Government ministers and leading industrialists and educators vie with each other to emphasise the crucial role for Design and Technology in the future, even for the survival of the national economy and for the long-term employment prospects for individuals. In Britain the Technical and Vocational Education Initiative (TVEI) began in 1986; The Training Agency (formerly the Manpower Services Commission) pumped millions of pounds into developing technological education in schools; millions more followed in the national extension. By the late 1980s, in some of the rural areas of England and Wales, British Schools Technology (BST) buses had become more frequent sights than passenger service buses. Meanwhile new GCSE syllabuses had rapidly translated the remnants of Craft, Design and Technology (CDT), Home Economics, and associated subjects into the new subject area. The National Curriculum with its incorporation of Technology as a foundation subject of which Design and Technology constitutes some 80 per cent of the total is speeding the changes still more rapidly in secondary and primary schools. Throughout the nation, there is evidence that, in primary and secondary schools, the new programmes are burgeoning. Shaw (1991) put the matter well:

When historians in the future look back to the early eighties and turn their minds to economics, the key word will be monetarism. When they turn their minds to education the key word will be pre-vocationalism. Unemployment in 1982 was running at 12.5% and inflation figures were still high: something had to be done. Ideological financial policies were matched by an ideological return to utilitarianism in education. Neither lasted long in their pure form but both were woven into the national fabric, and not only in the UK. By 1986 unemployment was down to 6% and much of the urgency had drained away. Cruder pre-vocational beliefs were becoming limited to YTS [Youth Training Schemes]. In schools TVEI [Technical and Vocational Education Initiative] flared up briefly, over a hundred different projects, but it was rapidly taken over by teachers despite the early intentions, and assimilated to the mainstream of education away from the narrower vocationalism and instrumentalism of the founders. TVEI reminded teachers of what was good in the non-traditional disciplines, of the importance of co-operation across boundaries and planning at the level of the institution, of the value of links with FE [Further Education] and of ways to motivate a wide range of pupils. Yet it may well be that TVEI will come to be seen as marginal in curriculum terms.

What survived all of this into the National Curriculum was Design Technology. It is the one conspicuous innovation in the Act [Education

Reform Act 1988], and it draws together elements from previous domains with an assurance that neither integrated science nor integrated humanities come near to attaining.

Yet although this book is about the newest area in the school curriculum it is also one of the oldest – the ways in which students in schools work with materials – with wood and clay, metal and concrete, paper and plastics, and with food and fabrics. But of course Design and Technology is not only about the manipulation of materials but also about the complex series of judgements, about the needs that give rise to that manipulation and the responses to such needs. So the evidence of Design and Technology activity is not only manipulation but also drawing, writing, talking and modelling.

The range of work is vast; the history of many of the activities is as old as humanity itself. Yet only recently have we begun to realise the full potential of this area of the school curriculum. Not only have we discovered a wide range of new and previously unused processes and materials, but we have also rediscovered the intellectual as well as the practical learning that can take place in work with materials. Above all we have realised more fundamentally than ever before that, in a modern society, human capacity to use and to modify the environment is critically determined by capacity to understand, plan and utilise resources of three-dimensional materials. Their availability and well-designed manipulation are as essential to the activities of an advanced industrial economy as they have been to those of any previous social system. Their importance in every phase of human pursuit from the most basic to the most esoteric is self-evident; their importance in the capacity of humans to express themselves has probably never been so great. And as the scarcity of natural resources intensifies and the cost of materials produced from them rises, the argument for work in the school that enhances thought and discrimination in their use becomes ever more compelling.

The task of this book is to show how, in recent years, this area of work has come to occupy a new and central place in the school curriculum, and in doing so has acquired a higher status and new identity and a new title: Design and Technology Education. The original areas of work with materials – woodwork, metalwork, home economics, craft, art and many others – form an important input to a new organic whole that emphasises the whole process of using materials, from preliminary analyses of need, through detailed design research, to the manufacture, testing and use of the product or service. It is a process in which the actual production of objects, though often of central importance, is but one part of the whole.

The book will define the new identity and show how it has been achieved. It will be illustrated with examples of practice in the schools, and will conclude with a consideration of possible future developments. But first it is necessary to explore some of the familiar roots of Design and Technology education, roots that are well known to almost every person who has been educated in the present century.

The traditional craft curriculum

Every British schoolboy, but few schoolgirls, once experienced the school workshop, that often isolated outpost of the school where the woodwork and metalwork teachers reigned supreme and where 'handicraft' was practised. The characteristic products of this area seemed to change little over the decades. The woodwork teacher prescribed exercises in planing and sawing and rewarded their successful performance with the opportunity to produce pipe racks and egg stands. Diligent application to the production of such objects was the way to the promised land of book racks, coffee tables and standard lamps. In metalwork the rituals were much the same; only the objects of devotion were different, worship of the potential of the wooden planes, saws and chisels being superseded by that of the files, hammers and hacksaws. When suitable deference to these ubiquitous tools had been achieved then other more powerful magic could be learnt: the soldering stove with its heavy copper soldering irons and their remarkable power to bond metals together; the brazing hearth with its capacity to reduce metals to nearly molten form; and above all the forge – a mass of glowing coals with its attendant tongs, anvil and heavy hammers. Ultimately, for those few to whom the innermost secrets could be revealed came the chance to learn the most powerful magic of all – the use of the lathe with its sophisticated range of processes: surfacing, boring, knurling, and even the capacity to produce screw threads by a procedure that came very close to automation.

For the young male in North America, life was much the same. Here his teacher was not a handicraft specialist but an industrial arts graduate. He taught in a 'shop' and usually offered a somewhat wider range of crafts including printing and 'electrics'. There were other minor differences – 'lumber' tended to come ready-machined, and for the American boy the joyous piles of wood shavings so characteristic of the British workshop were seldom achieved with such abundance. But in most characteristics the faith and its objects of worship were unmistakably similar.

For the girls, things were different. In school crafts, sex discrimination in the coeducational schools was rigid. In another separate and segregated area of the school, the female teachers of girls' crafts or domestic subjects undertook a wide range of quite different activities – cookery, needlework, dressmaking, childcare and home maintenance. In such areas other rituals, equally traditional and equally powerful, prevailed. The complex skills of pastrymaking, preparing a Yorkshire pudding, the baking of cakes and sponges of various levels of sophistication, the rules of family health and hygiene, the care of babies and young children, all had to be learnt as essential preludes to the role of the adult female. So too did the skills of 'good husbandry' ranging from the capacity to ultilise food scraps to the ability to pluck and dress a chicken and use the offal and bones for stock and the feathers for upholstery. There were also the dressmaking skills of fitting, altering, and even turning garments. All this and much else came within the wide ranging traditional span of the domestic subjects, which gradually developed under the successive names of 'domestic science', and subsequently 'home economics'.

But work with materials in the school was not confined to the so-called boys' and girls' crafts that take place in the woodwork, metalwork and home economics areas. There was – and is – also the art and design studio devoted to a wide range of two- and three-dimensional expression. Beyond the art room there are still further areas where work with materials took place actively and regularly in the school: a notable one is the field of applied science. In this area, where laboratories and workshops are linked together, a very great deal of work in which the specifically technological aspects of design are explored in detail occurs. And as one considers the other areas of the school curriculum, it is clear that activities with materials have a role in almost every other subject. Not only are they obviously relevant in geography, history and horticulture, but also in mathematics, language, and the whole range of 'academic' components of the curriculum. And on certain occasions in even the most traditional schools the work of the 'materials subjects' took on a dominant role in the life of the establishment as a whole. For the annual school play the woodwork teacher and his students literally moved on to the centre of the stage as they produced the often complex and ingenious scenery and props. Alternatively, on the occasion of the governors' meeting or the annual open day, the home economics teacher and her girls became – and often still become – visible to all as tea and scones are provided for the assembled throng of dignitaries. The acclaim which their activities predictably bring is an acclaim enjoyed and accepted by the school as a whole.

So far we have discussed work in a predominantly secondary school context. Yet practical work with materials is widespread in the primary school too. Limitations of school resources and of the physical strength and maturity of the pupils mean that the work may be rather more restricted in the range of processes and materials used. But there are still complex and wholly successful activities undertaken in many junior schools in producing original solutions to complex design problems, working models of great sophistication, three-dimensional sculpture of remarkable originality, and a highly competent range of craft products in clay, cane, paper, card and wood is well known and an established part of the curriculum. And the integrated project activity for which many of the schools are famous regularly and successfully involves Design and Technology.

The historical background

The introduction of practical subjects in the early elementary schools of Britain and North America was intimately associated with the utilitarian orientations that characterised 'education for the masses'. It was seen largely as a pre-vocational training for the rapidly growing ranks of manual industrial and domestic workers. This utilitarian justification was largely responsible for the introduction of the subjects in the 1880s in the British technical and trade schools, and later into the higher-grade schools of the 1890s and, in the twentieth century, into the senior forms of the elementary schools. Manual training and housecraft were included in the British Grammar school syllabus after 1902 often, again, on predominantly vocational grounds. Many of the pupils who entered these schools, right through until 1944, were likely to leave at minimum leaving age to enter the trades and other manual callings, their modest needs having been catered for. A detailed account of this historical period and the low-status problems that bedevilled it is to be found in Penfold (1988) who, quoting Sommerhoff, demonstrates how neatly this fitted with the preconceptions of society at large:

> Manual skills are no substitute for creative thought. It is our belief that the traditional kind of school metalworkshop can do more harm than good to the engineering profession. It fails to attract brighter children and it leaves children with the impression that becoming an engineer is the same as becoming a machine operator or mechanic. This, of course, is one of the common confusions which have brought the status of engineer so much lower in this country than in other industrial countries.

The problem has been exacerbated by a recurring feature of the Design and Technology subjects – their capacity to offer themselves, in an acceptable form, to even the least-motivated pupils. One of the special contributions of the subject area has been that it opens up a range of intellectual activity to many pupils who, because of their lack of verbal capacity, may be unable to take part in the highly verbalised intellectual processes of many areas of the school curriculum. The opportunity for non-verbal communication through this medium is important. It is an importance that has been recognised increasingly as we have come more clearly to realise the extent and consequences of the previously unnoticed verbal differences that effectively prevent communication between teachers and their students. Following the illumination of writers such as Labov (1972), many teachers are now wrestling with the task of finding alternative media of communication that will make the thought processes of many previously 'inarticulate' children available within the classroom.

It is seldom recognised how fully teachers of Design and Technology and their predecessors anticipated and solved such problems, not just through the use of 'non-standard English', but even through the ability to frequently avoid words in the communication of complex meanings. Perhaps it is only the headteachers who have regularly been aware of this special contribution and recognised it by offering the Design and Technology teacher large periods of time with the more difficult and less-articulate senior pupils and those in the lower streams of the secondary schools. Yet this achievement of the Design and Technology teachers has often only contributed still further to lowering their status by ensuring that their involvement with the more able and higher-status students was relatively diminished.

Faced with the persistent undervaluation of their work in a school system that has favoured intellectual achievement, many teachers have sought to 'intellectualise' their own subject. Most conspicuously this has taken place within the examination system, where a considerable effort was made to achieve the introduction of examinations in handicraft and domestic subjects by the various public examining boards. By the end of the old School Certificate system the craft subjects were recognised at both School Certificate and

Higher School Certificate level, and this recognition continued with the GCE examinations at Advanced and Ordinary levels in the late 1940s. Yet in becoming 'respectable' in this way the subject content of the crafts was put into a strait-jacket and the slow escape from the formalism of the nineteenth century was arrested. Up and down Britain the examination syllabuses in the craft subjects called for regurgitated pieces of memorised knowledge and skill from which almost all creative performance was eliminated. Students had to reproduce cross-sections of planes and micrometers, list and describe different kinds of wood screws and nuts and bolts, and perform rigidly prescribed practical set pieces in order to achieve their certificates. Yet despite such efforts the examination results in the craft subjects were received unenthusiastically by the academic world they were designed to impress and, with the dubious exception of technical drawing as a prerequisite for some university degree courses, students gained little from the possession of the certificates for which so much had been sacrificed. In professional training for craft and art teaching much the same kind of development occurred and the bizarre means adopted to give an intellectual legitimation to both the professional and in-service courses for craft teachers are well known. A convincing illustration of such courses is presented by Hanson (1971), who examines the strange procedures adopted by professional art teachers in their quest for professional recognition. Yet another example was the College of Handicraft with its academic correspondence courses and academic dress and rituals.

The turn of the tide

In the 1960s it became clear that, almost miraculously, the tide of recognition for Design and Technology was turning. It is difficult to single out any one major reason for the transmogrification. As with all changes in the curriculum, there are multiple causes, and it is impossible to apply precise weighting to any of them. Certainly there was a sequence of events that in no way applied solely to this area of education. The movement for curriculum renewal that flowered in the 1960s, carefully nurtured in Britain by the Schools Council, the Nuffield Foundation and other bodies and in the United States by a wide range of agencies with Federal support, has been exhaustively documented. At its most fundamental it sprang from the pressures of new knowledge and new understanding that was making obsolete much of what had been taught even in the recent past. The curriculum in mathematics had to be reviewed to take account of a computer age; developments in nuclear physics had transformed the content of even the junior science classes; work in written and spoken English had to take heed of the unprecedented development of new media of communication.

The characteristic responses to such pressures are well known. They included above all else a move to a curriculum that emphasised adaptability, originality and participation rather than memory, stored knowledge and passivity. No longer could the schools hope to provide an adequate stock of knowledge sufficient to last an individual through adult life, as had been envisaged in previous decades. The fact that knowledge was changing with ever-increasing speed made this unthinkable. Instead, the task of schools and particularly that of the Design and Technology teachers, was seen to be to help people to find and use new knowledge as it became available through their adult lives in work, home and leisure and, hopefully, to participate actively in the process of developing new knowledge. It was seen that the capacity to adapt, to initiate, to modify, to solve problems and to make decisions was likely to be the central human contribution to life in the coming decades, while the ability to reproduce knowledge or skills that had been so highly prized in the past was in future likely to be taken over by non-human devices. Teachers who doubted the need for the often repeated emphases on creativity, initiative and adaptability were able to find evidence that allayed their doubts in the falling demand for repetitive manual labour, coupled with the strident demand for individuals with initiative and ideas to be seen in the 'situations vacant' columns of the newspapers. There were also the urgent reminders of social economists such as Friedmann (1955) who wrote of the increasing need for most citizens of modern technological societies to change jobs at several stages during their adult lives, claiming that 'the myth of an individual with a single occupation value cannot be retained'.

The changing nature of schooling

Closely associated with these curriculum changes were others of equal relevance to craft educators. There was a new realisation that the very nature of experience in the school was changing. In an earlier age, students at school had been rich in first-hand experience of their communities. They were familiar with the way in which adults worked in the farm, in the factory and in the home. Essentially the curriculum was to provide them with the second-hand experience that was unavailable to them: the world of distant countries and of distant ages that could not come to them in any other way than through the words and skills of the teacher. But in contemporary industrial societies second-hand experience is available to the young as never before through television, film, newspapers and other illustrated printed matter. Yet the world of first-hand experience has for many become astonishingly limited. For young people living in a high-rise block and taking the bus to school it may well be that, apart from a few delivery workers bringing mail and milk, they have never seen an adult at work. Their parents may not wish to discuss the world of work on their return home in the evening. Increasingly, the school comes to be a place that organises and makes available first-hand experience and basic learning and consequently diminishes its concern with second-hand experience.

In all these new approaches, the Design and Technology subjects found themselves remarkably well placed to make an effective contribution: emphases on decision-making through the use of materials, the provision of first-hand experience, the integration of knowledge, the organisation of work in a community context – all these can be effectively provided by such teachers. A simple example was to be seen in a small school visited by the author. With his students, the technology teacher had set up a project where they designed and produced small pieces of equipment of use to the patients in a geriatric home near the school. The project was a classic exercise in Design and Technology decision-making. Students visited the old people in the home, talked with them, discovered their needs, returned to the school, tried out a range of designs on paper, returned over a period to the geriatric home, gradually produced prototypes, tested them in use, and finally made equipment that was of real value to the old people by making their lives somewhat easier. The objects produced included simple but well-designed aids for entering and leaving the bath, equipment for exercising, book supports to help the old people read with less fatigue, and much else.

But this was only part of the story. As the students talked with the old people the conversations took on a special significance. Previously there had been little or no contact between the two generations. As the old talked with the young both discovered new knowledge of the community in which they lived. The history of the community fifty to seventy years past was unknown to them and to most of their parents and was virtually unrecorded. With the support of the English teacher in the school a record of life in the community over half a century ago was undertaken. The active involvement of the history and geography teachers and of the staff as a whole followed. Here the work of one technology teacher had been central in determining not only the activities of a wide-ranging group of students but also a wide-ranging group of the teachers in the school.

The arrival of Design and Technology education

In the school mentioned above, and in many others up and down the country, the transformation of the materials curriculum has been achieved by the teachers and students themselves in their response to the new opportunities that have become available in the past decade. While still building upon the traditions of the past they have brought about the change that now allows us to define this area of the curriculum as Design and Technology education: a subject that is concerned with the identification and solution of problems in the use of materials that occur in the social systems in which our students will be adults. It is a subject area that involves a genuine fusion of intellectual and practical activities; its relevance to the contemporary world is self-evident; its appeal to students and teachers is unambiguous. Unlike its predecessors it has no need to resort to suspect devices to achieve academic legitimacy; its interconnectedness with all other activities of the school can allow us to demolish the barriers that for so long prevented the craft educator from achieving status and recognition.

The introduction of change

Many of the changes were speeded by two national projects which had reported in the 1970s. One was Project Technology charged with introducing technology to the schools, the other was the Design and Craft Education Project which aimed to transform the older craft areas of the curriculum to deliver a coherent design-based technological component. Both were financed by the Schools Council, the government-funded body which first began the deliberate total process of curriculum development in schools in England and Wales. The effect of the two projects was to create two separate reforming movements in the school curriculum. Project Technology sought to introduce a new subject into the curriculum that was additional to what had already been there. The Design and Craft Education Project sought to transform existing subjects to create the new design-oriented approach.

The 1970s were a period of dramatic change as both teams undertook extensive in-service work. Both also established influential magazines for teachers – Project Technology publishing *Schools Technology* and the Design and Craft Project publishing *Studies in Design Education Craft and Technology* which subsequently became *Design and Technology Teaching* (now the journal of the Design and Technology Association). Each project also published a string of widely-used books and pamphlets on ways in which new approaches could be put into practice – mostly written by teachers for teachers.

By the end of the 1970s, the time was appropriate for a further step forward. Schools had found ways to link Design and Technology and yet still had strong elements of traditional craft activity. Craft, Design and Technology (CDT) became the accepted name of the subject. Home Economics, Business Studies, and Art and Design, though remaining separate in most schools, became very closely related and were already becoming part of a Design and Technology system in many schools. By the 1980s CDT had become the official title and *The Sunday Times* (5 March 1989) was able to report that:

> Craft, Design and Technology (CDT) is the newest subject in the school curriculum, and unfamiliar to many people. But soon it will be as established as English or Maths – a compulsory element in every child's school experience.
>
> CDT covers a range of problem solving disciplines. Its clumsy name comes from the fact that it draws together many disparate elements. But because it is so all embracing it is one of the most important subjects for the 1990s.
>
> Included in CDT are skill based subjects such as woodwork, metalwork and working new materials such as plastics; technological elements such as electronics, which combine the practical and theoretical; and Design, which requires a mixture of intellectual skills, aesthetic awareness and practical experience of how materials behave.
>
> So CDT is much more than just a practical subject aimed at the less academic students who are 'good with their hands'. It involves a great deal of thought and planning bringing together manual, intellectual and organisational skills as no other subject does. It aims at preparing the student for living and working in a technological world.

The Assessment of Performance Unit

The next opportunity arose through the Assessment of Performance Unit (APU). This body was set up, within the Department of Education and Science, in 1975 to attempt to assess the performance of school pupils. It was built on the urgent necessity to record what the children were actually achieving in the new diversified curricula which had developed in the 1960s and the 1970s. Most projects – not only those in Design and Technology – paid relatively little attention to monitoring, and existing methods of assessment (ranging from school-leaving examinations to folk wisdom) were unable to assess much of the new aspects of children's learning that had been mapped out in the preceding decades.

But there is also another key consequence of the APU's work. By demonstrating the reality of achievement in the new aspects of the curriculum, it was able to reinforce the standing and esteem of the subject areas in which it worked. Conversely, subjects not in receipt of APU assessment were unable to enjoy this objective, public recognition of their capabilities.

The APU began its work in Mathematics, Science and English in primary and secondary schools. It later extended its remit to Modern Languages – but proposals for a wide range of other aesthetic and expressive subject groupings failed, largely because of the difficulties of assessment in those areas. However, a strong pressure group was established for the inclusion of Design and Technology and eventually a Working Group was set up by the APU. This, with strong HMI support, generated a pamphlet assessing *Design and Technology* which was published in 1979. This was highly influential, mapping out ways of assessing Design and Technology far removed from the prevailing school-leaving examinations of the time. The demand for the pamphlet was substantial; many thousands of copies were distributed.

By 1981 the APU had agreed to set up a feasibility study to explore the possibility of identifying and assessing Design and Technology performance. This was conducted at Trent Polytechnic and the results, published in 1985, seemed to indicate that there was 'something with a distinctive identity' that was not currently being assessed by the APU in its work in other subjects and it was therefore, potentially, capable of being examined by the Unit.

Eventually a major project was established at Goldsmiths' College, London, under the direction of Richard Kimbell. The work of the team has been devoted to identifying the distinctive capabilities developed by Design and Technology activity and also the identification of the teaching strategies wherein they may be optimised. The work of the Goldsmiths' Project is described in detail in Chapter 5 which is devoted to assessing Design and Technology but here it is important to mention that, like other APU subject explorations, the team soon found itself not only monitoring pupils' work but also identifying ways in which, through different teaching approaches and assessment strategies, their performance could be enhanced. Its recommendations, published in 1991, constitute a major document not only on assessing, but also on teaching, the subject. Its final report (1991) develops a major restatement of Design and Technology education:

> From the earliest work in this field, there has been general agreement on certain basic tenets of Design and Technology. It is an *active* study, involving the *purposeful* pursuit of a *task* to some form of *resolution* that results in *improvement* (for someone) in the made world. It is a study that is essentially procedural (i.e. deploying processes/ activities in pursuit of a task) and which uses knowledge and skills as a resource for action rather than regarding them as ends in themselves. The underlying drive behind the activity is one of improving some aspects of the made world, which starts when we see an opportunity to intervene and create something new or something better.
>
> All Design and Technology is essentially opportunistic in the sense that if we cannot see the opportunity, for example, to exploit or create a new market need, or recognise the opportunities inherent in this or that material or production technique, then the activity would never get underway at all. But assuming the activity does get underway – because we *do* see the opportunity to create something new or better – we have to recognise that the concept of 'better' is a problematic one.
>
> Whether you see something as being 'better' will

depend entirely on your value position. Is it 'better' to burn cheap fuels (based on hydrocarbons) or renewable fuels? Is it better to have whiter than white (i.e. chlorine bleached) nappies or duller ones? Is it better to have motorways, or the acres of open country that they use up? Inevitably Design and Technology impacts upon, and is influenced by the political, economic, physical and social world in which we live and these influences create the climate in which some outcomes are seen as more desirable than others. What is possible is not necessarily desirable – or at least it is not seen as desirable by all. Design and Technology is therefore at the cutting edge of social conscience where the concepts of 'need' and 'improvement' are far from clear and are often contentious.

Once inside a design task, value issues continue to be all pervasive, leading the designer to optimise one quality against another, prioritising one set of values against another, e.g. durability against cost, or visual styling against a particular material or production process. Design and Technology is unavoidably and continually concerned with identifying and reconciling conflicting human values.

The interaction of mind and hand

For APU we attempted to create a different way of looking at Design and Technology: a way that placed the interactive process at the heart of our work and the products as subservient to that process. To do this, we rejected the idea of describing the activity in terms of the products that result from it, and instead concentrated on the thinking and decision-making processes that result in these products. We were more interested in *why* and *how* pupils chose to do things than in *what* it was they chose to do. The pupil's thoughts and intentions were as important to us as were the products that resulted from them.

We gradually came to see the essence of Design and Technology as being the interaction of mind and hand – inside and outside the head. It involves *more* than conceptual understanding – but is dependent upon it, and it involves *more* than practical skill – but again is dependent upon it. In Design and Technology, ideas conceived in the mind need to be expressed in concrete form before they can be examined to see how useful they are.

The Technical and Vocational Education Initiative

Meanwhile much else was happening. In 1982 the Manpower Services Commission, an agency of the Department of Employment established, with ministerial backing, the Technical and Vocational Education Initiative (TVEI) which after piloting in schools in a limited range of Local Education Authorities spread to cover all secondary schools and many further education establishments. Its purpose was to offer a high level 'alternative' course in vocationally oriented technological studies – ranging from Computer Studies and Information Technology through to electronic music and photography – though most schools only offered a limited number of courses in this spectrum. Selected students commenced their TVEI courses at the age of 14 and either completed four years in their school or followed the last two years in a further education establishment – although by no means all completed a full four years.

A key element of the TVEI scheme was the enhanced funding received by the schools for equipment, materials and staffing. Very substantial sums were received by the first group of schools and this enabled them to make themselves ready to teach Technology in a way that could not have occurred without TVEI. In a real sense, TVEI has made it possible for National Curriculum Technology to be feasible in secondary education.

A further development has been the introduction of City Technical Colleges (CTCs) from 1985. This initiative, jointly funded by industry and government, has led to a small group of highly-funded schools in which the whole curriculum for all students is driven by technology. At the time of writing these colleges are few in number and their achievements uneven. Indeed an HMI report on the first CTC is highly critical of its technology teaching. However, government plans to increase CTC numbers substantially were presented in 1991.

The National Curriculum

By 1986 the prospect of a National Curriculum linked with legally-imposed pupil assessment had become a live political issue. There was widespread belief that the learning experiences of children were unacceptably uncertain and variable and that teacher effort and motivation were inconsistent. It was also believed that the light sampling approach adopted by the Assessment of Performance Unit was an insufficient national measurement of achievement. Above all, there was a general political will to control the work of teachers much more fully and in particular to make teachers and schools far more accountable to their clients – the parents and the communities they served. A consultation document on the National Curriculum was issued in July 1987 followed by legislation in 1988.

Just as it was important that Design and Technology was incorporated into the APU programme it was equally important that the subject became one of those identified as constituting the National Curriculum – one of the ten foundation subjects as they came to be called. Had the subject not been included, then its chances of survival, let alone the development of all that it offered, would have been modest. The pressure to incorporate it prevailed but in the legislation it became called 'Technology' – largely because of the view that the subject needed to incorporate more than Design and Technology – notably Information Technology.

The legislation for the National Curriculum was a major part of the Education Reform Act 1988. The passing of this Act was followed by the establishment of a Task Group on Assessment and Testing (producing the TGAT Report) which proposed a strategy of formal assessment strategies (Standard Attainment Tasks) and also recommended strategies for parallel assessment by teachers. It also underpinned the setting up of Working Groups in the various foundation subjects. Not surprisingly, Mathematics, English and Science were the first of the Working Groups to be set up.

There was much concern when there was a considerable delay before any announcement of a Working Group in Technology occurred. The concern was heightened when it was announced that a few additional members with technological expertise were to be added to the Science Working Group so that the Group could also report on primary school Technology. Many people feared that the powerful science lobby had achieved what amounted to a takeover of Technology – that it would become Applied Science and that Science would therefore effectively control 20 per cent of the curriculum rather than that 10 per cent specifically labelled 'Science'.

However Technology persisted. Under the redoubtable chairmanship of Lady Parkes, a vigorous Technology Working Group was established and produced a detailed interim report some six months after its establishment in 1988. Moreover, the group renamed itself the 'Design and Technology Working Group' and so titled its final report. This ambitious, wide-ranging document incorporated not only CDT activity but also Home Economics, Business Studies, Information Technology and aspects of Art and Design and generated its final report after a range of consultations in June 1989. This was referred by the Secretary of State for Education to the National Curriculum Council which published a consultative document in November 1989 with suggested amendments. At this point, the title reverted to 'Technology' – largely on the grounds that this was the title used in the legislation and could not be changed. However, the distinction in the final report between Design and Technology and Information Technology was maintained and with four of the five Attainment Targets in Design and Technology, the dismay of the Design and Technology enthusiasts was modest.

In the light of the National Curriculum Council's response the Secretary of State accepted virtually all the proposals of the final report of the Working Group and set the drafting of regulations in train. These were submitted to Parliament in August 1990 and published at the same time.

It is from this state of affairs that this book now proceeds – with Technology as an established Foundation Subject of the National Curriculum for all children aged 5–16. The full details follow in subsequent chapters. But this chapter has charted some of the long history which has led Design and Technology activity into the schools from its many humble beginnings into a position of high status, permanence and the opportunity to deliver all its benefits to children.

There is, however, still much to play for. The effective integration of work in Science and Technology is still to be worked out – both nationally with the subject associations and in each school with its still separate Science and Technology departments. Similarly the interface between Technology and Art & Design has also to be developed – the concept of Design in the school Art studio is often far removed from that in the Design and Technology department as the 1991 report of the Art Working Group makes clear. The case for an effective interrelationship with all other National Curriculum subjects is strong – see for example *Technology Education and its Bearing on English* (Medway, 1990). But this too will call for much finesse as our chapters on the National Curriculum on management make clear.

There is also a continuing need for Design and Technology teachers to be vigilant to avoid diminishing the subject. The problems being experienced by primary school teachers new to the subject may give rise to a case for reduction in the Technology component of primary educators. Simplified strategies for delivering the subject promoted by text book publishers (Eggleston, 1991) may dilute the necessary rigour. Moves to excuse some pupils from Technology from age 14 may again take able pupils away from the subject. Vigilance is essential!

Conclusion

This chapter has reviewed the development of Design and Technology from modest, low-status and diverse origins to a subject area that is important, significant and established. But development has been at a price – demanding major adjustments which have had to be undertaken at a rapid pace. The process of adjustment constitutes a theme of the remainder of this book.

References

Department of Education and Science (1988) *National Curriculum, Task Group on Assessment and Testing* (The TGAT Report). London: DES.

Eggleston, J. (1991) 'Pop goes Technology'. *Times Educational Supplement*, 13 December.

Friedmann, G. (1955) *Industrial Society*. New York: Free Press.

Hanson, D. (1971) 'The development of a Professional Association of Art Teachers'. *Studies in Design Education, Craft and Technology*, Vol. 3, No. 2.

Labov, W. (1972) 'The logic of non-standard English', in P. P. Gigliol (ed.) *Language and Social Context*. Harmondsworth: Penguin.

Medway, P. (1990) *Technology Education and its Bearing on English*. Leeds: Leeds University School of Education.

Penfold, J. (1988) *Craft Design and Technology: Past, Present and Future*. Stoke-on-Trent: Trentham.

Schools Examinations and Assessment Council (1991) *The Assessment of Performance in Design and Technology* (The Goldsmiths' Report). London: HMSO.

Shaw, K. E. (1991) *Teaching Design Technology* (Perspectives 43). Exeter: Exeter School of Education.

CHAPTER 2

What is Design and Technology education?

This chapter explores the distinct yet linked natures of both Technology and Design and considers why both are essential components of the school curriculum. Examples of work in school are offered and the implications for adult roles in work, home, leisure and community are discussed.

As we have seen in Chapter 1, Technology is now a compulsory subject from the age of five for all children in State schools in England and Wales up to 16 years. Moreover, there is every sign that many independent schools are following suit. Yet the precise identity of this new subject is still unclear to many teachers – either through total or partial unfamiliarity with it in their professional training or in their experience to date. This chapter attempts to remedy this deficiency for teachers and, hopefully, to enable them to know and to explain the nature of their subject to even more bewildered parents, employers and pupils.

Design and Technology is unique in the school curriculum. It is the one subject directly concerned with the individual's capacity to design and make, to solve problems with the use of materials and to understand the significance of Technology.

It will be easiest to begin to define Design and Technology by reference to Technology in the National Curriculum. Essentially this is defined in the Attainment Targets. These are:

- *Attainment target 1: Identifying needs and opportunities*
  Pupils should be able to identify and state clearly needs and opportunities for design and techno- logical activities through investigation of the contexts of home, school, recreation, community, business and industry.
- *Attainment target 2: Generating a design*
  Pupils should be able to generate a design specification, explore ideas to produce a design proposal and develop it into a realistic, appropriate and achievable design.
- *Attainment target 3: Planning and making*
  Pupils should be able to make artefacts, systems and environments, preparing and working to a plan and identifying, managing and using appropriate resources, including knowledge and processes.
- *Attainment target 4: Evaluating*
  Pupils should be able to develop, communicate and act upon an evaluation of the processes, products and effects of their design and technological activities and of those of others, including those from other times and cultures.

Some teachers have seen this as a 'design and make' process but it is much more than that as the Final Working Group Report (DES, 1989) and the ensuing Regulations and exemplary material (DES, 1990) make clear. These documents will be referenced and explored in detail in Chapter 3.

Defining Design and Technology

Although the National Curriculum provides a useful working basis, it cannot, and does not, attempt to be the authoritative and complete definition of Design and Technology. Indeed, there is no doubt that the Working Group would not have accepted this responsibility even if it had been asked to exercise it. At the simplest level Design and Technology has two components – 'Design' and 'Technology' in close relationship. It consists in using Technology to achieve solutions that satisfy sound design criteria and using design to achieve solutions that satisfy sound technological criteria. As we have seen, most secondary schools have, for a very long time, offered a range of technology and design activity in areas such as food, fashion, work with wood and metal, applied science, business education and much else. The importance of this relationship and integration is strongly emphasised in the Working Group Report (1989) sections 1.20 and 1.21 which read:

1.20   The activities of design and of technology overlap considerably. As we said in the Interim Report, 'most, though not all, design activities will generally include technology and most technology activities will include design'. However, we believe that the core of knowledge and skills as encompassed by our programmes of study, taken alongside the four attainment targets we detail, cover the significant aspects of design.

1.21   That is not to say that the knowledge, skills, values and processes of designing cannot be used and developed in other subjects. For example, in environmental design pupils will rarely be involved in creating a totally new environment, but will need to appraise what already exists, explore needs and devise ways of organising and achieving change. In pursuing their ideas, they will develop their sense of historical and cultural continuity and a recognition that the new has to grow out of the old. There are clear opportunities here for work in history, but also for other subjects such as geography, to build on and develop these ideas.

The close integration of Design and Technology is emphasised by statutory orders for National Curriculum Technology. They require that, at each Key Stage pupils' Design and Technological capability is to be developed through:

- A broad range of practical activities. In each key stage, pupils should design and make: *artefacts* (objects made by people); *systems* (sets of objects or activities which together perform a task); *environments* (surroundings made, or developed, by people) in response to needs and opportunities identified by them.
- Five broad contexts of work (situations in which design and technological activities take place): home, school, recreation, community, business and industry. Work should progress from familiar to unfamiliar contexts.
- Working with a broad range of materials including textiles, graphic media (such as paint, paper, photographs), construction materials (such as clay, wood, plastic, metal), and food.
- A breadth of knowledge, skills, understanding, attitudes and values required in the attainment targets and programmes of study. In addition, 'pupils should be taught to draw on their knowledge and skills in other subjects, particularly the foundation subjects of science, mathematics and art, to support their designing and making activities'.
- Personal development through activities in Design and Technology. 'Pupils should be taught to discuss their ideas, plans and progress with each other and should work individually and in groups. . . . They should be taught to take reasonable care at all times for the safety of themselves and others . . . Activities should also reflect their growing understanding of the needs and beliefs of other people and cultures, now and in the past.'
- Progression of individual capability. 'As pupils progress, they should be given more opportunities to identify their own tasks for activity, and should use their knowledge and skills to make products which are more complex, or satisfy more demanding needs.'

The integration of Design and Technology is interestingly expressed in a course module at Middlesex University. Headed *The Technology Dimension* it asserts:

> The principal objective of this module is to enable students to develop their knowledge and understanding of the Technology dimension in the past, at the present and in the future and in a variety of cultural contexts. It will seek to challenge and develop students' critical awareness and understanding of the influence of designers/technologists in all sectors. The module has three key components.
>
> *Images of Progress – The history of style and fashion.* Design alters the way people see artefacts, systems and environments. Capitalism depends on this capacity to innovate and sell products thus creating wealth. This component will explore how style and fashion have and are being created and how they are exploited.
>
> *Engineering our Environment – the history of Technology.* Engineering, in all its guises, has been responsible for improving every aspect of our lives and will continue to do so in the future. Every new innovation brings benefits for some and problems for others. This component will explore the impact and implications of technological innovation.
>
> *The Design Revolution – the impact of design on industrial strategy and performance.* The impact of new technologies, the blossoming of international competition and global markets, the tension created between the creative function of design, goal orientated marketing and profit taking investment are all aspects which this component will explore. The theoretical analysis of issues such as product life cycle, technological maturity, globalisation etc. will be set in the contexts of case studies drawn from a range of activities and scales. This will include contact with design professionals from all elements of the design spectrum.

But even though there is abundant evidence of integration, it remains true that few schools have presented the curriculum in a way that fully incorporates it or even identifies the true nature of Design and Technology. In very many schools Technology and Design are still seen as lesser areas of activity – taking place in workshops and studios, respectively, with a dominantly practical nature and largely unrelated to the other subjects in the curriculum.

The incidence of Technology issues across the curriculum was interestingly displayed in an early Assessment of Performance Unit study undertaken at Trent Polytechnic. It divided Technology into three components, value judgements, knowledge and skills and explored where they occurred in the curriculum. The results are indicated in Fig. 2.1. The exercise threw up some interesting illumination – for instance the considerable attention being paid by Religious Education classes to the technology of warfare and its human and social consequences (the study was undertaken during the Falklands War and could almost certainly have been replicated during the Gulf War).

The importance of values in any full study of Technology has been emphasised by Ruth Conway (1990):

> . . . technology education is not just an instrumental activity, giving pupils the knowledge, skills and resources to be able to make things, but should be encouraging pupils to harness their creative abilities towards goals that they have consciously chosen and evaluated, with growing sensitivity to the needs of other people and the environment, and responsible decision-making.

As with Technology so the incidence of Design in the curriculum is broadly based. There are obvious, direct examples in the work of the Art and Design, CDT and Home Economics areas but there are other examples across the curriculum. Computer programs, environmental design in the sciences, the presentation of English, environmental design issues in social studies – the production of a play in drama, the scoring of a composition in music – there are many examples. Indeed, there is a Design element in the delivery of virtually every lesson in school and in every pupil's response. Without it they would lack form, focus and the capacity to be evaluated. As with Technology, Design is mistakenly seen to be concerned only with three-dimensional activity on many occasions – the error is fundamental and highly misleading.

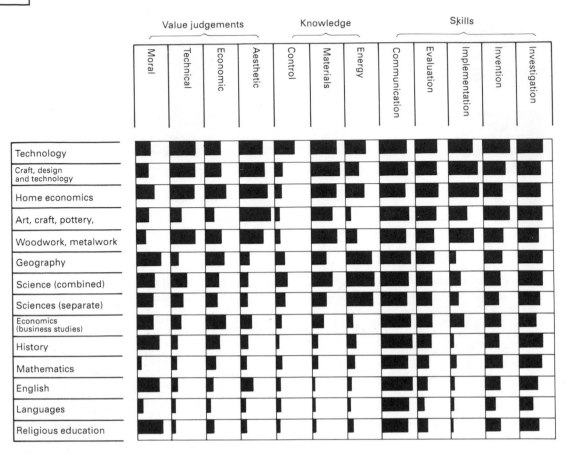

Fig. 2.1   The contributions of subjects to technological understanding (Trent Polytechnic Survey for APU, 1983)

The nature of Technology

The problems of identity of both Technology and Design in the curriculum are exacerbated by the long and complex history of both areas in society at large. Technology has played a major role in the development of civilisation – in the long path from primitive, nomadic, minimal livelihood to the opportunities of sophisticated modern society. Yet until comparatively recent times the technologists who created the cathedrals and all other enduring buildings, the roads and highways, the canals, the coaches and other vehicles, the heating and lighting systems and much more, have been largely invisible. Usually only the patrons were identified. Only in the eighteenth and nineteenth centuries did things begin to change spectacularly with the achievements of some of the technologists – Wren, Brunel, Arkwright and others. It was only in the nineteenth and twentieth centuries that the great national institutions were founded: The Institution of Mechanical Engineers, The Institution of Chemical Engineers, The Royal Institute of British Architects, The Institute of Chartered Accountants – all these were nineteenth or twentieth century inventions. Indeed, it is only in very recent times that accountants and bankers have been recognised as technologists in any regular way. Only in the mid-twentieth century has the accreditation of Technology and associated programmes of study been fully formulated for virtually all branches of technology.

Elsewhere Technology was being delivered by craftsmen and women or people who, in the twentieth century, have become more generally known as 'technicians'. The immense social significance of the distinction between the terms technologist and technician cannot be overemphasised. The term 'technician' came to cover the full range of people who *did* technology – who made things work and continued to make them work. But their status was inferior; they worked to the orders of their patrons, clients, employers or managers who were not required to have any of the requisite practical technical knowledge or capability. The establishment of the role of professional technologist such as the engineers, architects, town planners and financiers reinforced rather than diminished this distinction. In their training and in their professional work there was no need for them to lay bricks or saw timber. No mechanical engineer needed to be able to machine metal or assemble machinery. Essentially their role was cerebral rather than manual.

The segregation of technical education followed the segregation of the occupational structure. The most able children in the schools – selected by their achievements in academic subjects (notably Science, Literature and Language) – proceeded to higher education and to follow professional and managerial courses. The less able left school at minimum learning age and, unqualified, became apprentices or some form of on-the-job trainees, perhaps to achieve technician status. For some, training became available in Technical Schools and Vocational Classes. We have already considered, in Chapter 1, the low status of these courses and the low status prospects of those who followed them. Essentially they were seen as avenues for less able pupils to become occupationally useful.

Technology education in the National Curriculum marks an attempt to override such old distinctions and especially to break out of the low status of technical education and to bring technological education from higher education into the schools. It is also an attempt to demonstrate that technology is an appropriate and important subject for the education of all children including the most

able. Furthermore, it is making the point that not only technicians but also technologists and indeed all citizens need to be able to understand, develop and handle technology in all its aspects. If it succeeds it may achieve a major goal – to break down the status barriers which have so long impeded the economic development of England and Wales and many other Western countries – the low status of actually making things in a system controlled by those who do not. The spectacular success of Japanese manufacturing industry where, culturally, these divisions do not exist offers a justification of the need for such a change. But to achieve these goals it is vital that Technology in schools does not suffer the same pitfalls and devalue sound practical capability. The danger of turning technology into a subject where realisation is only the making of drawings and models is considerable.

But of course Technology education is not only about occupation. Every citizen needs to be familiar with a wide range of technology in order to have sufficient understanding and capacity to live effectively in modern society: electrical, financial, child-rearing and architectural technology – these and many more technologies determine the quality of life and range of opportunity of every citizen. Individuals must not only be able to involve themselves in technology but also be able to enter into effective dialogue with the professional technologists who every day are making key decisions about their lives and welfare. Just as the technologists need to be able to think and make, so do the non-technologists need to be able to make and think about technology.

The nature of Design

The nature and status of Design is at least as complex as that of Technology. Archer (1973) writes:

> Design is that area of human experience, skill and knowledge which is concerned with man's ability to mould his environment, to suit his material and spiritual needs. . . . There is a sufficient body of knowledge for this area called 'design' to be

developed to a level which will merit scholarly regard for the future.

At the heart of the matter is the design process. This is the process of problem-solving which begins with a detailed preliminary identification of a problem and a diagnosis of needs that have to be met by a solution, and goes through a series of stages in which various solutions are conceived, explored and evaluated until an optimum answer is found that appears to satisfy the necessary criteria as fully as possible within the limits and opportunities available. The design process at its most complete is one that can be used to describe, to analyse and hopefully to improve every aspect of human activity and especially those human activities that lead to end products and services. Jones (1970) puts it effectively when he says 'the effect of designing is to initiate change in man-made things'. But of fundamental importance in the concept of design is rationality. The design process above all else is one of rational, logical analysis. Jones emphasises this strongly, commenting that the picture of the designer is

> very much that of a human computer, a person who operates only on the information that is fed to him and who follows through a planned sequence of analytical, synthetic and evaluative steps and cycles until he recognises the best of all possible solutions.

Defined in this way, the concept of design was given its most powerful impact in the work of the Bauhaus, an industrial arts school in pre-war Germany. Here, in close association with artists and thinkers such as Gropius, Kandinsky and Klee, there developed a new and powerful movement to explore fundamentally and rigorously the process of design as a human activity.

Until the time of the Bauhaus the form of hand and machine-made objects had normally been achieved by a combination of tradition, expediency and chance. Design was commonly a unilateral activity in which the requirements of one participant tended to predominate and often to monopolise the specifications. Thus, the craftsman alone could impose considerations of skill or availability of material; the engineer alone could impose technological requirements; the client alone could impose considerations of taste or finance. Not infrequently the result of such 'designing' was brilliantly successful – but only occasionally and incompletely was it rational. The Bauhaus set out to change all this. Students were encouraged to study the process of design in a way that was both total and detailed. The results were of central importance; the new wave of industrial design that began in the inter-war years revolutionised the chaos of design in a multitude of manufactured products. To some extent, many of the most famous industrial products of the mid-twentieth century owe some debt to the influence of the Bauhaus – including the Braun food mixer, the Volkswagen 'Beetle' and the Olivetti typewriter.

In the art colleges of Europe and the United States, the Bauhaus influence was widespread. It did much to develop among students a concern for purity and simplicity of form and an appreciation of properties of materials, of colour and texture, that by comparison with what had gone before appeared to be austere, even chaste. To a great extent it was responsible for the concept of the foundation course, still almost a mandatory part of most art school courses, in which students undertake fundamental explorations of the nature and the property of materials. But it is easy for any new movement to become obsessional, and the rationality and purity of the design movement was no exception. De Sausmarez (1964) wrote

> basic design is in danger of creating for itself a frighteningly consistent and entirely self-sufficient form, a deadly new academicism of general obstruction for young painters and young designers – a quick route to the slick sophistication of up to the minute graphic design.

Like any system of ideas, the design process was certainly guilty of over-sophistication, rigidity and abuse. A particular problem was that design came to be an exercise for designers. It was an exercise in which, by virtue of their knowledge of the rules, they came to hold power and control, and in which all other participants, often including the client, came to be imprisoned in the designer's ethic.

More recently, we have come to realise yet again that there are many participants in the process of design and that not all of them act in a wholly rational way or even accept the 'rationale' of the designer.

In the closing years of the present century there has been a sharp reaction to the design movement, as individuals have sought to reimpose their feeling and individuality on designed products of all kinds led, in the UK, by Prince Charles in his commentary on modern architecture. At a personal level, this is perhaps most strikingly to be seen in the world of leisure where, for example, motor-car owners have sought, with much energy, to make their mass-produced cars distinctive from those of other people, and thereby in some way to express their own self-image and life style. It is only necessary to purchase a copy of a popular car magazine or to read the works of Tom Wolfe to see how widespread and effective such a movement is. It is a movement that will sometimes lead individuals to apparently extreme lengths to satisfy its aims. Fashions in leisure equipment such as motor cycles, clothing and music systems may be adopted not so much on technological criteria but on their potential for self-image, style and personal expression – as any parent of adolescents or even young children knows well.

The design and technological process in the schools

The preceding sections alert us to a realisation that Design and Technology is a shared activity between those who make things and those who use them – an activity in which very many people participate, ranging from those who plan and execute manufactured products to those who acquire an object at second or even third hand. It is an understanding that is now widely appreciated by all who design, whether they be product managers, artists, architects, landscape specialists or consumers. It is this socially-sensitive concept of technology and design that is at the heart of Design and Technology education as it is developing in the schools. Below there are reproduced diagrams of

the design process and the technology process adopted respectively by the Schools Council Project on Design and Craft Education (Fig. 2.2) and Project Technology (Fig. 2.3). Both diagrams illustrate clearly the similar detailed and analytical process of enquiry that leads to the achievement of Design and Technology and the meaningful social context and range of participation within which the process is undertaken.

The experience of Design and Technology education

What are the practical consequences of the above activities within the school? Let us consider how the design and technology process might be experienced by a group of students in Key Stage 4 of a local secondary school. Let us suppose that the school has a project that aims to involve its students more fully in the community in which they live, to give them the experience of responsibility, decision-making and participation in local affairs. Let us assume that the school is serving a large and somewhat underprivileged housing estate in an industrial city. As in many similar situations, the school is attempting to improve the facilities of the neighbourhood as part of the project. An often needed facility is equipment for younger children to play with in their leisure time. The designing of such equipment would almost certainly be seen to be a major responsibility of the Design and Technology department of the school. How would this be done? The teachers and their students would already have a preliminary knowledge of the problem and an understanding of the human purposes at issue. These would, however, require some further exploration. How large is the relevant child population of the area? What is its age distribution? Is the birth rate rising, falling or remaining steady? Information of this kind, which could be sought from the Education and Social Services Departments of the local authority, would give some indication of the demand for facilities and the existence of any play facilities in the neighbourhood which, unknown to the school, were in fact being used. Are there special reasons, such as traffic hazard or violence, that would make

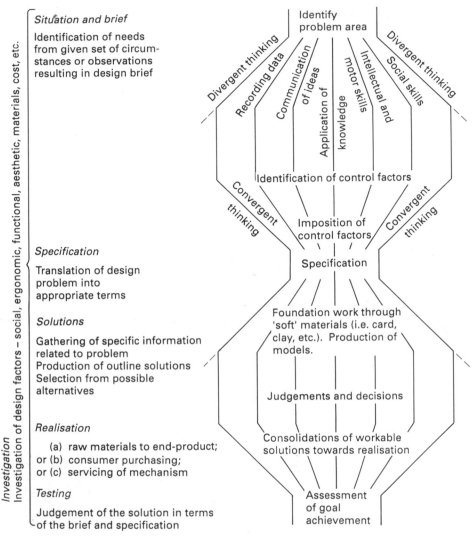

Fig. 2.2   Design process – Design and Craft Education Project (from *Design for Today*)

mothers of young children unwilling to use such facilities even though they did exist? Questions such as these may call for survey work by the students in and around their own homes, for visits to the parents of young children, the police and other civic authorities. In this way, a precise profile of the nature and extent of the need for play facilities may be reached.

With such a profile compiled, further prelimi-

nary work would be called for in which the constraints and resources available could be considered in detail. As with all design processes it is important to explore the experience of the past:

● Were there previous attempts to provide play facilities and, if so, why did they fail?
● Are there problems in their design that, with hindsight, could be remedied or are there in-

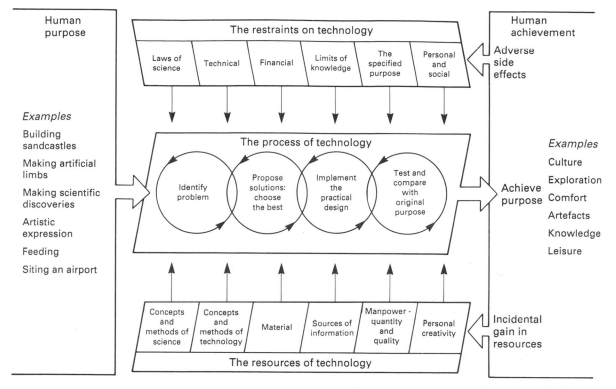

Fig. 2.3   The process of technology – Project Technology

herent difficulties that call for an entirely new approach?

- What of the likely location of the play facilities? Are they to be indoor or outdoor or both? If indoor, who will provide the premises? If outdoor, who will provide the land and ensure that it is suitably enclosed?
- Are the locations available feasible from the point of view of access or do they require parents to bring their children long distances or risk crossing busy main routes?
- Would the location be supervised when in use and, if so, by whom? Would this be undertaken by senior students at the school or would there be regulations that required the presence of at least one trained adult? If the latter, who would be responsible for the recruitment and possible payment of such an adult?

- If supervision was essential could the location be effectively closed off from use when such supervision was not provided?
- Would it be necessary to ensure supporting facilities such as first-aid equipment and toilets?

All these in-depth questions concerning constraints and resources are essential preliminaries to the designing and construction of the play equipment itself. Only if the answers to such questions are satisfactory is there any prospect that the equipment, once provided, can and will be used in the manner envisaged. All of the enquiries are within the scope of co-operating groups of students and teachers. All of them have a relevance far beyond the specific issue at stake – a relevance that is likely to have meaning in the later lives of the students in this or other communities.

Having undertaken this still preliminary work, detailed consideration of the facilities to be provided may be undertaken. Some kinds of facility will already have been readily eliminated such as, for example, soft toys if the location is to be an outdoor one. The next step is to consider precisely what will be provided. Let us assume that the preliminary enquiries have failed to throw up the possibility of an indoor location or adult supervision, but that a sector of a local park is to be made available and surrounded by 'child-proof' fencing by the park authority. It is, however, unlikely to be 'vandal-proof'. In such a situation, the equipment to be constructed will need to be permanent, non-portable, and capable of withstanding weather and possible physical violence as well as normal usage by the young children.

Narrowed down in this way, the design process can take on a sharper focus:

● What sort of equipment satisfying these criteria do young children enjoy? Students may well visit play facilities beyond the community or in schools where they are provided and study children at play, noticing what equipment is popular and the use to which it is put. Detailed measurements may be taken not only of the equipment but also perhaps of the sizes of the children themselves.

● What are the financial resources available? Can the school raise funds to purchase suitable supplies of, say, metal tubing to produce climbing equipment; hardwood, new or second hand, to provide benches; concrete tubes to make tunnels and obstacles? Possibly the local authority will have funds available for this purpose or a community association may make a donation, or it may even be possible to arrange a sale of work or jumble to raise funds.

● What of the skills available? Does the teacher have competence in the use of concrete? If not, is it possible to attend a course run by the Concrete Development Association, or can assistance be obtained from a local builder or contractor who may even be able to arrange for the loan of a concrete-mixer for a short period?

● Is welding equipment available for the construction of tubular climbing frames? Can the school's circular saw cope with thick sections of hardwood?

● What contribution can the Art teachers make concerning the colour combination and the aesthetics generally of the proposed equipment?

● Can the Home Economics trained teachers advise, from knowledge of the physical development of young children, on the more beneficial kinds of equipment that may be provided?

● And outside the Design and Technology department, to what extent can the other departments of the school – Social Studies, Language, Mathematics and Science – provide resources both of understanding and expertise? Can the project benefit these other departments in turn? Is there perhaps the opportunity for an interesting study by the English students of the way in which language is used during play in the formulation of rules?

Considerations of this kind will have led to a specification of objects of play equipment to be constructed within the Design and Technology Department that take into account as fully as possible the needs and capacities of the users, the resources and competencies of the school and its personnel, the nature and properties of the materials envisaged, and perhaps most importantly, the creative capacity of the participants to devise new solutions that go beyond modifications of existing ones. With such specification, possible solutions may now be explored. Mock-ups of envisaged equipment can be constructed; children of the appropriate age-group can be invited to try them out and their responses recorded; the equipment can be modified and rearranged on site, and the responses compared with those to previous arrangements; the aesthetic consequences of various groupings of equipment can be appraised. This part of the process may occupy an extended period of intense activity as step-by-step modification and improvements to the original solutions are formulated.

Meanwhile, there may be further detailed consideration of the resources of the school for production:

- Do the workshops or studios need modification?
- Is new material or equipment to be ordered and assembled?
- Are production-line arrangements needed, and if so how can they best be planned in order not only to ensure sound production arrangements but also to give students the opportunity of experiencing how such arrangements may be optimised through rational discussion?
- Are there likely to be bottlenecks in certain aspects of the production? If so, how may they be eliminated or at least minimised?
- How may individuals be trained to undertake specific tasks with which they may be unfamiliar? Is a rigid division of labour likely to be the best solution, or will some members of the production group become bored by being involved in more repetitive tasks? If so, does there need to be some system of job rotation?
- How will quality control be maintained? Will there be need for safety testing of the individual components of the swings, climbing frames and other facilities before they are finally assembled? In all these considerations the underlying need for sound basic craft capability is ever present.

Eventually, when the design is finalised and the production arrangements are confirmed, the actual manufacture of the equipment can begin. But the process is by no means over, for not only has the manufacture to be completed skilfully, but the equipment has also to be installed effectively, and possibly some of the construction work has to take place on the actual location itself. It is then necessary to ensure that the facilities are satisfactory in use, to undertake post-delivery checks and maintenance over a period. Ultimately, the knowledge gained in the post-production period becomes the raw material for further Design and Technology processes in the future. All of this final stage would constitute the evaluation required in Attainment Target No. 4 (see p. 32). Taken together the whole process would embrace all four Attainment Targets at Levels of Attainment 4–10 and would allow them to be assessed by the Standard Attainment Tasks and Teacher Assessments. However, it is important to emphasise that not all aspects of a full design and technology process have been mentioned in this example. Notably, the whole question of market research, sales promotion, marketing and accounting, items that form essential components of the design process in a normal industrial company, have not had to be taken account of fully in this non-commercial example. But in suitable projects these activities too should form a part of the total process of Design and Technology education experienced by students; they too are important components of the industrialised society in which they will spend their adult lives and in which enterprise awareness is an essential requirement. In describing such a project we have been portraying activity which might form a Long Standard Attainment Task (SAT) of the kind discussed in Chapter 4.

Some characteristics of the Design and Technology experience

This example of the working out of the Design and Technology process in a secondary schools shows many of the features of the design process in action. Readers will be able to draw up their own list of activities that are taking place. Among others it is important to notice the following:

1 Students have had the opportunity to experience and participate actively in an inventive and creative process in which new ideas can be developed and old ones modified. In these processes they have also had the experience of responsible decision-making in which not only have their ideas been used, but the responsibility for this use and its consequences on the lives of others has been unmistakably their own.
2 Students have come to see, in a way often impeded by traditional school curricula, the interplay of knowledge and understanding, how the work of one 'subject' complements and augments that of another, and how few, if any, problems can be solved with a narrow subject orientation. Many ideas, many resources, and many materials are called for in the solution of almost all human problems.

3 Students have become aware of the social context of human behaviour. This example, like almost all other manifestations of the design and technology process, made it clear that decision-making cannot be undertaken in isolation. The resources and needs of the clients, the parents, the community as a whole, even the wider society, have to be taken into account, and solutions have to satisfy all these participants if they are to be, in any real sense, an adequate response to a problem.

4 The long-standing concerns for skilled performance and the integrity and honesty of workmanship are all honoured in a project of this kind. Skilled work is essential if the equipment constructed is to serve the purpose for which it was intended, and the safety of its users to be ensured. In addition other materials such as concrete, requiring different but still unmistakably skilled handling, may also be introduced. The difference from many previous craft activities is that here skill is being used in a meaningful rather than an artificial context: its acquisition and its employment can be justified to even the most cynical parent or the least sympathetic teacher.

5 Above all, the design and technology process provides a tool of enquiry which, once experienced, will probably have a wide general applicability in the adult life likely to be experienced by the students; it is a process that will link rather than isolate them from the economic and social aspects of adult community in which they live.

In reviewing the consequences of the Design and Technology process in school we have returned once again to its social justifications, especially those that concern the key decision-making elements and the social integration to which it leads.

Applications of Design and Technology education

Occupation

We have already referred to the changing nature of work in technological society. In the past, education was based upon the assumption that children were 'stocked up' in school with knowledge, values and skills that would stand them in good stead throughout their adult life. The object was to enable them to become, as near as possible, like the present generation of adults; the ultimate mark of attainment was to attain adult standards in the use of existing knowledge and skills. Modern technology has rendered the traditional memorising of knowledge and skills of very much less importance; computerisation, the development of data banks and retrieval systems, make the human memory a less precious commodity; automation and production lines make replicated hand skills less necessary and even inappropriate for an increasing range of artefacts. Yet this is not to deny the continuing importance of the experience of highly skilled performance, the ability to make connection by memory and recall. It is these capabilities that allow humans to use technology and to develop and enhance it successfully.

Even where hand skills may still be regarded as appropriate for production the cost of labour-intensive human production may make them unacceptable. In a range of industries from the Post Office to the railways the high cost of labour leads to increasingly thoroughgoing attempts to modify or to minimise production methods that involve a high input of human labour. Increasingly, young people no longer find themselves in traditional apprenticeships, or in semi-skilled and unskilled manual roles. While still handling materials, they may be handling them in ways that, in a sense, suggest the business executive far more than the traditional craftsman. The future of an increasing number of the children who leave our schools will lie in the servicing industries, maintaining motor cars, office equipment, domestic and leisure apparatus of all kinds. Here much of the work should certainly be highly skilled but it will also involve a range of other tasks including diagnosis, deciding between replacement, repair or destruction, the responsibility for maintaining time sheets, the collection of money, the organisation of a work schedule, the running of a service vehicle and much else.

Even young people who enter industry in an apparently traditional occupation will find that work may be different: they may enter the local engineering factory to become store keepers and may well find that the work no longer requires the careful memorisation of myriad different patterns of nuts, bolts, screws and washers. Instead, they may find that the stores they are handling are valves, transistors and other pieces of electronic equipment. Visually they are almost indistinguishable from one another, so their nature has to be recorded not in visual appearance but through a system of numbering and colour coding that may change frequently. The store keepers have to learn a system which they can develop and adapt to the changing nature of commodities. Commonly it will be a system linked to a computer that will undertake a great deal of the labour-intensive work of keeping stock records, ordering new supplies, and the associated costing and accounting. Again, the store keeper's need is to become adaptable, even inventive, and above all to be capable of making rational decisions based upon evidence. The relevance of the Design and Technology process to the needs of late twentieth-century occupations is unmistakable; it may well be argued that it would be irresponsible of the schools not to provide such experience.

Home, leisure and community

What is true in work is arguably true in leisure and in the home. The opportunities to express oneself through one's choice of materials as a consumer, as a skilled gardener, boat builder or motor-racing enthusiast have never been greater. The economic and technological resources of modern society all combine to make this possible by often generous rates of pay, particularly to young people, and also in the provision of longer hours of leisure. But it is important to realise here that, as with the world of work, we are defining roles in words very different from those in which we have traditionally defined them. In the past we have spoken not only about workers but about artists, woodworkers, needle-workers and the like. But here we are talking about the generic roles that increasingly characterise modern society – such as home owners, sportsmen and women and amateur gardeners: designations that imply a wide-ranging use of ideas and materials. The integration that is central to the Design and Technology process – the integration of specialist skills in a social context – is also central to the roles of modern society.

But there is yet a further and possibly even more compelling characteristic of the Design and Technology education component of the school curriculum. It springs from the new realisation in modern societies on both sides of the Atlantic that the material environment, both public and private, is above all a product of countless individual decisions. For many years we were convinced that decisions about our material environment were best left to the experts. Accordingly, we trained small numbers of highly selected designers, town planners, town and country planning officers, landscape artists and the like, and believed that, with their training, these specialists would be able to make appropriate, wise decisions. All that then remained was to persuade the majority of people, largely through the experience of schooling, to respond to their wisdom. In the Design and Technology subjects in particular, we endeavoured to introduce the experts' decision to our students and to encourage them to accept them. We took them to the Design Centre and other exhibitions of 'good design' and allowed them to see the products which they could and should use in their homes. We took them to see the exhibitions of the planning consultants for their city and taught them to respect the wise suggestions made for the development of their civic environment. In doing this we overlooked, as did many of the adherents of the Bauhaus (see above), the realisation that the environment is never wholly determined by the decision of planners or consultants, even though indeed their suggestions are at times adopted. It is determined far more by the way in which ordinary men and women use their environment. It is determined by what they plant and what they construct in their gardens, by the way in which they decorate and locate their caravans, by the way they use public facilities and by the way they decide to spend their money and their leisure. A leading consultant on urban planning recently conceded, after many years of successful and well-regarded public practice, that he has now found it necessary to take into account the fact that, regardless of almost every known control, ordinary people found ways of constructing sheds or out-houses in their back gardens. He announced that henceforth he would 'take cognisance' of this in the preparation of his schemes!

In such ways we have become aware that we are more likely to achieve 'good design' in our environment if we recognise the participatory nature of design processes. Nicholson (1972) has gone so far as to suggest in his 'theory of loose parts' that the more successfully that technologists and designers create a 'non-participant environment', the more successfully will people attempt to participate and establish their individual presence in it, even to the extent of behaviour which is labelled as vandalism. The 'structural modifications' that take place in waiting rooms and public conveniences help to make Nicholson's point. Increasingly an education designed to inculcate respect, to put people into a 'received' environment, is gradually giving way to an education in which people are expected to participate in decision-making processes concerning their environment. In doing so they become active rather than passive participants in a modern society.

Conclusion

The experience of Design and Technology education that has been described in this chapter is nothing less than a preparation for such a participative role, and it is argued that this constitutes the most sufficient justification for the new subject's identity of design education. Hudson (1966) has noted that 'Our education should create an environment where an individual can discover something of himself, his aptitudes, the relevance of his ideas and of other people's ideas.' The task of subsequent chapters will be to identify more fully the practice of this essential preparation for adult life offered by Design and Technology education.

References

Archer, L. B. (1973) 'The need for design education'. Paper presented to DES conference N805, Horncastle. Mimeo. London: Royal College of Art.

Assessment of Performance Unit (APU) (1983) *Design and Technology Performance (Trent Polytechnic Survey)*. London: DES.

Conway, R. (1990) 'Values in technology'. *Times Educational Supplement*, 28 September, p. 22.

Department of Education and Science (1989) *Design and Technology for Ages 5–16* (Final Report of the Working Group on Design and Technology). London: HMSO.

Department of Education and Science (1990) *Technology in the National Curriculum*. London: HMSO.

De Sausmarez, M. (1964) *Basic Design: The Dynamics of Visual Forum*. London: Studio Vista.

Design and Craft Education Project (1971) *Design for Today*. London: Edward Arnold.

Hudson, T. (1966) 'Creativity and anti-art', in *Design Education*. Hornsey: Hornsey College of Art.

Jones, J. C. (1970) *Design Methods and Technology: Seeds of Human Futures*. London: John Wiley.

Nicholson, S. (1972) 'The theory of educational sloyd'. *Studies in Design Education, Craft and Technology*, Vol. 3, No. 2.

Project Technology (1971) *Final Report*. London: Schools Council.

CHAPTER 3

Design and Technology in the National Curriculum

This chapter reviews the place of Design and Technology in the National Curriculum for England and Wales. It identifies the constraints and the opportunities provided by the legislation and some of the ways in which teachers may respond to them. Though largely specific to England and Wales the events described have many parallels in other countries and other educational systems.

We have described the long, complex and often difficult path that has led Design and Technology to its established place in the National Curriculum so that, with Information Technology, it accounts for about 10 per cent of the National Curriculum and is a mandatory foundation subject for pupils from 5–16 years. Even allowing for the range of permitted variations at 14–16 (Key Stage 4) and possible changes between 5–11 (Key Stages 1 and 2) it shows every sign of being a permanent feature of the National Curriculum for all children for almost all of their schooling.

Although the arrival of National Curriculum Design and Technology has been widely welcomed by those who teach it, realisation of the sheer magnitude of what has now to be delivered has led teachers to undertake fundamental reappraisals of what they have taught in the past. No group of teachers has experienced this problem more acutely than those in primary schools (Key Stages 1 and 2, ages 5–11 years). Their training for teaching has seldom given serious attention to Design and Technology and most have had very little experience of specific Technology activity in their classrooms. Yet, paradoxically many primary school teachers have quickly made impressive developments in teaching the subject unencumbered by all the long-standing traditions and impediments of the previous subject forms which many of their secondary school colleagues have lived through. Indeed the contrast between the approach of primary school teachers (new to the subject and with the National Curriculum requirements as their starting point) and secondary school teachers (with a diverse and often long-standing set of earlier approaches to Design and Technology and adjusting to the new requirements) soon became one of the major features of the development of Design and Technology. The transfer of pupils who have experienced Design and Technology from the age of five to secondary school at age 11 promises to be much more dramatic than the change from Key Stage 2 to Key Stage 3 in many aspects of the subject.

What does the National Curriculum require?

The requirements of National Curriculum Design and Technology are common to all students – primary and secondary. What are these requirements? The best guides are, of course, the statutory regulations and the accompanying non-statutory examples. However, this chapter will

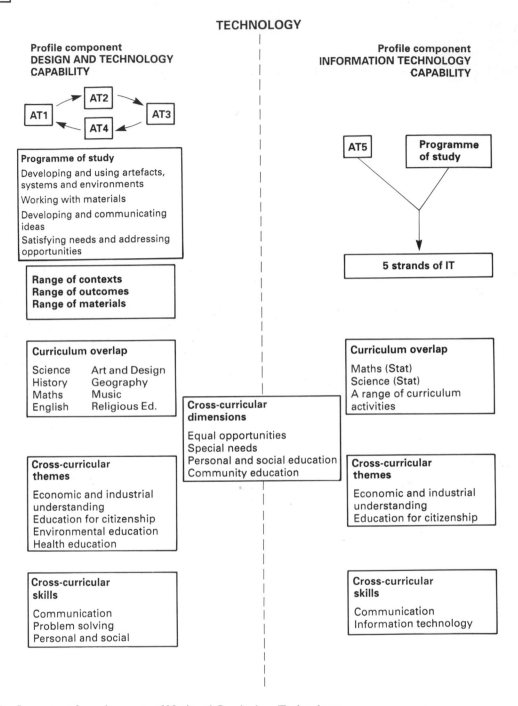

Fig. 3.1   Summary of requirements of National Curriculum Technology

review the main dimensions which will form a useful prelude to the exhaustive, detailed information contained in the *Statutory Orders* partly reproduced in the document *Technology in the National Curriculum* (DES, 1990). In it we shall have to use much National Curriculum official jargon but we shall try not to make it too oppressive!

Design and Technology capability, to give it its full title, is one part of National Curriculum Technology. The other part is, of course, Information Technology capability which is the subject of another book in this series. The overall structure is portrayed in Fig. 3.1.

In Chapters 1 and 2 we have charted the route whereby the subject with its Working Group on Design and Technology progressed through an interim report in 1988 (DES, 1988), a final report in June 1989 (DES, 1989) followed by a National Curriculum Council Consultation Report in November 1989 (NCC, 1989) and Statutory Orders in March 1990 (DES, 1990).

The Approach of the Working Group

The Working Group made an early declaration of approach which received widespread support. It asserted:

- the development of Design and Technology capability 'to operate effectively and creatively in the man made world' as the overall objective for the subject
- contexts for design and technological activity which are broad, balanced and relevant
- within the attainment target framework, the co-ordination of design and technological activities currently undertaken in art and design, business studies, CDT, home economics and IT
- the use of knowledge, skills and understanding drawn from the core subjects of Mathematics, Science and English
- attainment targets which reflect the holistic nature of Design and Technology
- the description in programmes of study of a core of knowledge, skills and values as resources to be used in design and technological activity

In June 1989 the Group prefaced its strategy for attainment targets and programmes of study as follows:

> . . . we have aimed to ensure that they provide the means by which pupils develop the ability:
>
> - to intervene purposefully to bring about and control change
> - to speculate on possibilities for modified or new artefacts, systems and environments
> - to model what is required in the mind, symbolically, graphically and in three-dimensional forms
> - to plan effective ways of proceeding and to organise appropriate resources
> - to achieve outcomes of good quality which have been well appraised at each stage of their development
> - to appraise artefacts, systems and environments created by others
> - to understand the significance of design and technology to the economy and to the quality of life

The Attainment Targets

At the heart of the proposals are the four Attainment Targets (originally five) that identify Design and Technology. We have already used them to help to define the subject in Chapter 2 but for convenience they are reproduced here.

- *AT1 Identifying needs and opportunities*
  Pupils should be able to identify and state clearly needs and opportunities for design and technological activities through investigation of the contexts of home, school, recreation, community, business and industry.
- *AT2 Generating a design*
  Pupils should be able to generate a design specification, explore ideas to produce a design proposal and develop it into a realistic, appropriate and achievable design.
- *AT3 Planning and making*
  Pupils should be able to make artefacts, systems and environments, preparing and working to a plan and identifying, managing and using appropriate resources, including knowledge and processes.

- *AT4 Evaluating*

  Pupils should be able to develop, communicate and act upon an evaluation of the processes, products and effects of their design and technological activities and of those of others, including those from other times and cultures.

For Design and Technology capability the four Attainment Targets, together, can possibly be seen as a design-and-make process. But to say this is to oversimplify, for in the old 'design approach' of CDT the main assessment of achievement was in the finished product – the summation of the design process. In Design and Technology capability all of the parts are valued in their own right.

The targets for each of the Attainment Targets are spelt out in 10 graded Statements of Attainment, Levels 1–10. Thus there are 40 statements of Attainment for Design and Technology capability and 10 for Information Technology capability.

It is these Statements of Attainment that are assessed by the Standard Attainment Tasks and the Teachers' Assessments which are described in Chapter 4. Specifying the various levels was a major task for the Working Group. It reports (p. 9):

> Specifying levels of attainment is far from being a simple matter of expanding knowledge and skills incrementally. In some instances, levels are increased by extending the range of performance, such as working with a broader range of resources or working in unfamiliar contexts. In other instances, progression is the result of a more sophisticated use of familiar resources or a deeper exploration of a familiar context. Progression also involves an increased interplay of knowledge and skills, value judgements and personal qualities.
>
> Another feature of progression is the ability to reflect upon practice and from this make explicit the concepts, procedures and strategies involved so that these can be carried over and applied consciously to new design and technological situations.

The best way to portray the Statements of Attainment is to offer a small sample together with the non-statutory examples given in the official document (see Table 3.1). Progression through these levels is crucial.

Table 3.1 Examples of Statements of Attainment

Statements of Attainment	Examples
Attainment Target 1 **Level 2** Pupils should be able to:	
● Describe what they have observed, visualised and found out in their exploration.	*Describe different methods to create movement in pop-up toys and books.*
● Suggest practical changes that could be made in response to a need and describe to others why they suggested certain changes.	*Suggest reorganising the home corner/practical area in the classroom so that toys can be stored more effectively.*
● Ask questions which help them to identify needs and opportunities for design and technological activity.	*Find out how the school cook chooses the menus for school dinners.*
Level 9	
● Demonstrate how they have devised and implemented a strategy for the investigation of unfamiliar situations which draws on their previous experience of design and technology.	*Prior to work observation at a shop, consider relevant knowledge and skills acquired and identify aspects of shop work which they might explore further and how this might be achieved.*

Table 3.1—continued

Statements of Attainment	Examples
Attainment Target 2 **Level 1** Pupils should be able to:	
• Express their ideas about what they might do to meet an identified need or opportunity.	*Draw picture showing different ways of scaring birds in a field of crops.*
• Report the progress of their ideas showing how they have clarified and developed them.	*Produce a series or set of drawing showing how the design developed, with details of drawings, models, plans, patterns.*
• Extend their first ideas by combining various aspects of them to formulate a design proposal and explain why some ideas were not used.	*Combine their proposals for fabric, colour, style and cost of toddlers' clothing to make a marketable product.*
• Seek out and organise information to help them develop their ideas and refine their design proposal.	*Use magazines, encyclopedias, databases, videos, etc. to make informed choices about the range of kitchen surfaces, storage spaces, appliances available when designing a kitchen.*
• Establish and check the availability of the resources required, adapting their design as appropriate.	*Check time, materials, skills, tools and equipment required and adapt their design in the light of these constraints.*
• Specify what they intend to do and what they will need by using simple plans and flow diagram.	*Draw up a plan for an automatic greenhouse watering system, and listing what they need, including information, materials, equipment, skills.*
Attainment Target 3 **Level 6** Pupils should be able to:	
• Plan and organise making in order to achieve the desired outcome.	*Use flow charts, prepare equipment.*
• Combine knowledge of the properties of a range of materials and processes, and identify those most suitable for design.	*In making a piece of jewellery, take into account qualities such as durability and the malleability of different parts of the construction and the way the material will need to be worked.*
• Demonstrate, by their choice and use of a variety of tools and equipment, that they understand the limitations of them and the need for safety and accuracy.	*Develop a simple jig to enable work to be done quickly, safely and more accurately.*
• Use knowledge of materials, components, tools, equipment and processes, to change working procedures to overcome obstacles as making proceeds.	*Make modifications to take account of a cost increase of a component or the discovery of a more appropriate alternative. Make simple modifications to the control program of a piece of equipment.*
• Show judgement in seeking advice and information.	*While designing a page of the school magazine, seek expert advice regarding the equipment and processes available to them.*
• Use knowledge of technical and symbolic representations of materials, components and processes to assist making.	*Use drawings and plans to assist making.*

Table 3.1—continued

Statements of Attainment	*Examples*
Level 10 Pupils should be able to: ● Use a range of techniques, processes, resources with confidence, safety and creativity to achieve high-quality work. ● Review the design proposal during planning and making and show resourcefulness and adaptability in modifying the design in the light of constraints to make a high-quality product.	*Use a combination of computer-aided design and other high-quality graphic techniques to produce a house style and image for a new company.*
Attainment Target 4 **Level 1** Pupils should be able to: ● Describe to others what they have done and how well they have done it. ● Describe to others what they like and dislike about familiar artefacts, systems or environments.	*Describe how well they made a mask and whether it fitted well and was strong enough.* *Describe what they like about their school bag.*
Level 7 ● Present an evaluation of their activities against the original need, drawing on information gathered about product and the reactions of users. Evaluation should include suggestions for improvements.	*Evaluate an aid for old people in terms of consumer response to the product, and its cost-effectiveness. Explain value for money, effectiveness in use, style and fashion.*

Key Stages

The four Key Stages are a prime determinant of the levels for which children are taught. The identification of Key Stages is a combination of children's ages and the composition of the school, classes or teaching groups. The guidelines are specified in the Appendix to Circular 3/90 (DES, 1990):

> An individual pupil in a teaching group may of course be younger or older than the age of the majority. Section 3(6) of the Act [Education Reform Act 1988] makes clear that it is the teaching group which a pupil is in for each foundation subject, not the registration class, which determines the key stage which is applicable to the pupil for that subject. So a pupil could be taught with another age group for one or more subjects where appropriate, e.g. in order to pursue that subject at a higher or lower level, while being taught with his or her peer group for other subjects. There is

nothing, however, in the Act to require pupils to repeat a year, or to move early to a higher year group.

It is envisaged that a pupil's passage through the levels will normally be linked to the Key Stages. This may be shown simply as follows:

● Key Stage 1 normally ages 5–7, Levels 1–3
● Key Stage 2 normally ages 7–11, Levels 2–5
● Key Stage 3 normally ages 11–14, Levels 3–7
● Key Stage 4 normally ages 14–16, Levels 4–10

Thus it is expected that by the age of 7 all children will at least have reached level 1, by 11, Level 2 and by 14 level 3. However, the notes to Circular 3/90 emphasise flexibility and suggest most children should be capable of achieving around Level 2 by age 7, Level 4 by age 11, around Levels 5–6 by 14 and around Levels 6–7 by 16. The overlap is intended to allow for different rates of progress by individual children and also to ensure continuity between stages.

Each Statement of Attainment is specified by regulation and is mandatory – schools must ensure that it is taught appropriately. However, as our illustrations have shown, each statement is accompanied by illustrative examples which are not part of the legislation. It must also be emphasised that the levels do not have to be taught sequentially; the teacher's task is to ensure that all children in each Key Stage achieve the highest levels they are properly capable of reaching. So that a child who can reach higher levels without working through lower levels may do so; similarly, a child may 'drop back' to find an appropriate level at which to succeed. There are of course, exemptions and alternatives for children with special needs and problems. The Statutory Orders define them as:

Pupils with special educational needs
The programmes of study and attainment targets are intended to cover the great majority of pupils, particularly when due allowances have been made as indicated in the Statutory Order:

Pupils unable to communicate by speech, writing or drawing may use other means including the use of technology or symbols as alternatives.

and

A pupil who, because of a disability, is unable to undertake a practical activity required under the programmes of study, may undertake an alternative activity which most closely matches that activity.
                                              (HMI, 1991)

For severely handicapped children a Statement may be made (Section 18, Education Reform Act):

. . . a statement made under the 1981 Act may modify or disapply any or all of the requirements of the National Curriculum if they are inappropriate for an individual pupil. The connection with a statement ensures that any departure from the National Curriculum will be decided in the light of educational, medical, psychological and other evidence about the pupil, including the views of the pupil's parents.

However, schools are required to ensure that all pupils have maximum possible access to the National Curriculum as the HMI Report (1991) makes clear.

Programmes of Study

The formulation of the actual Programmes of Study as opposed to the targets and levels was also a major task for the Working Group as they sought to ensure an open-ended, pupil-centred approach. The Group originally recommended 16 components; the Statutory Orders reduced these to four:

- developing and using artefacts, systems and environments
- working with materials
- developing and communicating ideas
- satisfying needs and addressing opportunities

A further change was that the Programmes, originally built around the 10 levels were changed to focus on the four Key Stages. *Technology in the National Curriculum* (DES, 1990) offers the following general guidance:

In each key stage pupils should design and make:

- artefacts (objects made by people);
- systems (sets of objects or activities which together perform a task); and
- environments (surroundings made, or developed, by people);

in response to needs and opportunities identified by them.

Contexts (situations in which design and technological activity takes place) should include the home, school, recreation, community, business and industry, beginning with those which are most familiar to pupils, and progressing to contexts which are less familiar.

Pupils should be taught to draw on their knowledge and skills in other subjects, particularly the foundation subjects of science, mathematics and art, to support their designing and making activities. These activities should also reflect their growing understanding of the needs and beliefs of other people and cultures, now and in the past.

As pupils progress, they should be given more opportunities to identify their own tasks for activity, and should use their knowledge and skills to make products which are more complex, or satisfy more demanding needs.

Pupils should be taught to take reasonable care at all times for the safety of themselves and of others.

Pupils should be taught to discuss their ideas, plans and progress with each other, and should work individually and in groups.

At each key stage pupils should be given opportunities to work with a range of materials, including textiles, graphic media (such as paint, paper, photographs), construction materials (such as clay, wood, plastic, metal), and food.

A pupil who, because of a disability, is unable to undertake a practical activity required under the programmes of study, may undertake an alternative activity which most closely matches that activity.

Throughout these programmes of study, the term materials includes components, and the term equipment includes tools.

The requirement concerning contexts for design and technological activities is reinforced at the start of the programmes of study for each key stage, as shown below:

Key Stage 1
Pupils should develop design and technology capability by exploring familiar situations (such as home, school and local shops). They should also look at familiar things (such as pictures, poems, stories, television programmes) as starting points for some of their design and technological activities.

Key Stage 2
Within the general requirements of design and technology, activities should encourage the appraisal of artefacts, systems and environments made by others as well as the application of enterprise and initiative.

Key Stage 3
Within the general requirements of design and technology, pupils should have increasing opportunities for more open-ended research, leading to the identification of tasks for designing and making. There should be opportunities for some of these activities to take place outside school.

Key Stage 4
Activities should include at least one extended design and technological task, for example with a duration of between 15 and 30 hours. There should be opportunities for visits and work outside school, including work experience placements.

Again it is possible to illustrate the Programmes of Study together with the non-statutory examples (see Table 3.2).

Table 3.2 Examples of Programmes of Study

Programmes of study	Examples
Key Stage 1 Levels 1–3	
Developing and using artefacts, systems and environments	
Pupils should be taught to:	
● Know that a system is made of related parts which are combined for a purpose.	*A bicycle; a house.*
● Identify the jobs done by parts of a system.	*Bicycle chain; a kitchen.*
● Give a sequence of instructions to produce a desired result.	*Prepare a shopping list in order of shops to be visited.*
● Recognise, and make models of, simple structures around them.	*Making a model building from simple construction kits.*
● Use sources of energy to make things move.	*Stretched elastic bands to turn a propeller on a model plane, a battery to make a toy move; moving things manually.*
● Identify what should be done and ways in which work should be organised.	*Stamping a pattern on a fabric.*

Table 3.2—continued

Programmes of study	Examples

Working with materials
Pupils should be taught to:

- Explore and use a variety of materials to design and make things.
- Recognise that materials are processed in order to change or control their properties.
- Recognise that many materials are available and have different characteristics which make them appropriate for different tasks.
- Join materials and components in simple ways.
- Use materials and equipment safely.

Use a variety of materials such as cotton reels or building blocks to make a tower; making a collage.
Yeast dough to bread; clay to pot.

Fabric, paper, card, clay, paint, wood, clay for making a beaker; newspaper for covering the table when painting.

Gluing card, sewing on buttons.

Developing and communicating ideas
Pupils should be taught to:

- Use imagination, and their own experiences, to generate and explore ideas.
- Represent and develop ideas by drawings, models, talking, writing, working with materials.
- Find out, sort, store and present information for use in designing and making.

By brainstorming, role-play, drawing, painting, modelling.
Draw plans of possible layouts, of a home for the pet they have or would like.

Satisfying needs and addressing opportunities
Pupils should be taught to:

- Know that goods are bought, sold and advertised.
- Realise that resources are limited, and choices must be made.
- Evaluate their finished work against the original intention.

Talking about what shops sell.
Sharing materials provided for a task.

Does the model car move as intended?

Key Stage 2   Levels 2–5

Developing and using artefacts, systems and environments
Pupils should be taught to:

- Organise and plan their work carefully, introducing new ideas, so that their work improves.
- Allocate time and other resources effectively throughout the activity.
- Control the use of energy to meet design needs.

- Use a variety of energy devices.
- Plan how practical activities may be organised.

- Use a variety of information sources in developing their proposals.
- Use knowledge and judgement to make decisions in the light of priorities or constraints.
- Identify the parts of a system and their functions, and use this knowledge to inform their designing and making activities.

Working as a group to produce a puppet play; making a garden or a nature area.
Setting up a school shop, making a school newspaper.

Making something move; using switches, taps or valves to switch something in their product on or off.

Making three-colour block prints; building up a fabric collage.
Books, videos, databases, other people.

Choosing the menu for a party, given a spending limit.

Triangular frameworks in pylons; departments in a supermarket.

Table 3.2—continued

Programmes of study	Examples
Working with materials Pupils should be taught to: ● Use equipment safely. ● Select materials for their task. ● Rearrange materials to change their strength or character, and to increase their usefulness. ● Join materials in semi-permanent forms. ● Assemble materials. ● Avoid wastage of materials.	 *Using criteria such as cost, availability, purpose, weight.* *Folding and bending paper; adding thickening agents to dyes and paints.* *Gluing card and wood to make a buggy frame.* *String and wood to make a bow.*
Working with materials Pupils should be taught to: ● Take responsibility for safe working. ● Develop co-ordination and control in using equipment. ● Finish work carefully.	 *Agreeing and following class rules for safe working.* *Painting details on a puppet's head; tidying threads on a knitted garment.*
Developing and communicating ideas Pupils should be taught to: ● Use modelling to explore design and technological ideas. ● Use modelling and recording when generating ideas. ● Break design tasks into sub-tasks and focus on each in turn as a way of developing ideas. ● Use materials and equipment to produce results which are aesthetically pleasing.	 *Making a preliminary model kit of card, timber, clay.* *When designing a model fairground roundabout, consider the means of movement, the type of structure and the appearance; when designing a turtle graphics program for drawing a row of houses, plan, write and test it in separate procedures.*
Satisfying needs and addressing opportunities Pupils should be taught to: ● Know that the needs and preferences of consumers influence the design and production of goods and services. ● Recognise the importance of consumer choice and hence the importance of product quality and cost. ● Be aware that the appearance of artefacts and environments is important to consumers and users. ● Know that human shape, scale, proportion and movement affect the forms of designs. ● Understand that goods may be designed to be produced singly or in quantity, and that this affects what each item costs. ● Consider the possible consequences of their design proposals before taking them forward to completion. ● Consider the needs and values of individuals and of groups, from a variety of backgrounds and cultures.	 *Discussing school meal preferences, fashion, clothes for hot and cold weather.* *Making a database of household goods and exploring it to find the best value.* *Carrying out a survey of people's views on the visual appeal of a game they intend to make.* *Furniture, telephones, pots and pans, toys.* *Comparing the costs of designer clothes and chain store clothes, handmade and mass-produced furniture.* *Will it be safe? How will it affect others?* *Planning a recreation area for both young and old; making food associated with a religious celebration.*

Table 3.2—continued

Programmes of study	Examples

- Evaluate at each stage of their work.
- Make adjustments as a result of evaluation.
- Use their appraisal of the work to help their own work.

Looking at pottery, painters' use of colour, packaged goods.

Key Stage 3   Levels 3–7
Developing and using artefacts, systems and environments
Pupils should be taught to:

- Analyse the task and its components to identify those which depend upon the completion of previous tasks and to develop a flow chart.

When planning a meal, identifying the courses and the items/preparation needed; making a three-colour batik.

- Set objectives and identify resources and constraints.
- Organise their working to complete the task on time.
- Produce a documented plan for their work, including an analysis of the resources required and a time schedule.
- Select and use mechanisms to bring about changes and control movement.

Mechanisms such as linkages and gearing; changes such as direction of motion or speed.

- Know that using energy affects comfort and convenience.

Heating, lighting, sound, air conditioning.

- Use information sources in developing their proposals.

Book, database.

- Analyse a system to determine its effectiveness and suggest improvements.

Supermarket check-outs; road traffic layouts; arrangements for school meals.

- Test simple objects to determine performance.

Test items of clothing for waterproofing quality; finding the maximum load for a carrier bag.

Working with materials
Pupils should be taught to:

- Ensure that the working area is ordered and safe, and that equipment is well maintained.
- Use equipment safely; follow safe working practices and understand the procedures for dealing with accidents.

How to identify what to do; whom to contact.

- Consider, when selecting and using materials, their physical and aesthetic properties, availability and cost, and the product being made.

Making a musical instrument; making a kite; using a material which will not rust when making a rudder for a model boat.

- Combine materials to create others with enhanced properties.

Making a sauce; mixing glue; using sand to give strength to clay; using interfacing to strengthen fabric.

- Assemble a range of materials.
- Take account of the constraints imposed by equipment.

Be aware of how much data they may reasonably expect a database program to handle in computer memory.

- Know the working properties of a range of materials.
- Recognise the purpose of equipment, to understand the way it works, and to use it.
- Identify hazards in the working environment and to take appropriate action if dangerous situations occur.

Trailing electrical leads; hot or sharp equipment left dangerously; bad design of work area.

Table 3.2—continued

Programmes of study	Examples

Developing and communicating ideas
Pupils should be taught to:

- Use specialist vocabulary when communicating proposals.

  Use words like 'file', 'record' and 'field', correctly communicating design plans to the purchaser; advertising information to a prospective purchaser.

- Develop styles of visual communication which take account of what is to be conveyed, the audience and the medium to be used.
- Present their design and technological ideas and proposals using modelling techniques and specialist vocabulary.
- Recognise the relationships between two-dimensional representation and three-dimensional forms.
- Investigate artefacts, systems and environments to find ideas for new designs.

  Tipper lorries, bridges, Saxon jewellery, Greek theatres.

Satisfying needs and addressing opportunities
Pupils should be taught to:

- Identify markets for goods and services.

  Parents needing refreshments at a school fête.

- Know that, in the production and distribution of goods, the control of stock is important.

  Investigating sales of ice-cream in the summer and winter.

- Plan a simple budget.

  A simple spreadsheet to estimate costs and income.

- Investigate the effects of design and technological activity on the environment.

  Motorway construction; flooding of valleys; landscaping of a derelict site.

- Establish and apply criteria for assessing the needs and opportunities identified; the choice of material and equipment to achieve the design; the procedures adopted; the end result.

Key Stage 4   Levels 4–10
Developing and using artefacts, systems and environments
Pupils should be taught to:

- Prepare a flow chart and a detailed work plan to achieve the objectives of the design.
- Use information sources in developing their proposals.

  Use anthropometric data in designing a chair.

- Allocate tasks when leading a team.
- Estimate the operating costs of a system, its dependency on other systems, and evaluate its efficiency.

  The cost of gas and electricity for central heating systems.

- Reduce energy loss and understand why this is important.

  In the home, office, industry, transport.

- Recognise that forces of different types are involved in structures.

  Tension, bending.

- Maximise the efficiency of a mechanism.

Table 3.2—continued

Programmes of study	Examples

Working with materials
Pupils should be taught to:

- Use equipment safely.
- Know that organisations need to have procedures for health and safety, and people responsible for enforcing them.
- Join materials in permanent forms.
- Have a working knowledge of the properties of a range of materials.

- Use materials economically and efficiently.

- Develop test procedures, including those for quality control.
- Know the properties and operational characteristics of a range of components.
- Develop and apply understanding and knowledge of how materials are shaped, cast, joined and formed.
- Understand that equipment can be adapted to serve a variety of purposes.

- Use computer systems in designing and making.

- Give attention to detail and accuracy.

- Develop craft skills.

- Know the efficient mechanisms depend on the appropriate choice of materials used and the number, form and arrangement of their component parts.

Examples (right column):

Rules for safe movement in school; laws for the movement of dangerous waste; health and safety officers; fire protection officers.
Soldering, brazing.
Specific shrinking of clay when fired; the residual memory of plastic; absorption of dyes in fabrics; melting points of alloys; corrosion resistance of metals.
Design the production of an object that just exceeds the specification.
Test shear on a garment.

Resistors, capacitors.

Joining dissimilar materials such as wood and acrylic; vacuum forming; casting aluminium.
Computer-controlled knitting machines, graphics programs, desk-top publishing; making a model lift, car park barrier or burglar alarm controlled by computer.
Three-dimensional images on a plotter; three-dimensional objects on a lathe.
Completing a book illustration; completing a presentation model or finished garment; ensuring that components fit.
By taking time to develop understanding of and to respond to the materials, equipment and processes being used.

Developing and communicating ideas
Pupils should be taught to:

- Analyse alternative solutions to produce a better design proposal.
- Design the appearance of an artefact, system or environment so that it appeals to users.
- Use modelling techniques to communicate design proposals.
- Use intuition as well as empirical data in developing their design.

Consider alternative means for warning when a refrigerator door has been left open.
Designing a radio for teenagers, furniture for adults.

Prototypes, garment models, projection drawings, organisation charts.

Satisfying needs and addressing opportunities
Pupils should be taught to:

- Develop a product and how to market, promote and sell it.

Designing and producing an entry system for a disco, car alarm; a healthy snack food.

Table 3.2—continued

Programmes of study	Examples
• Investigate ways in which solutions could be extended to meet additional needs.	*Converting a wind-powered pump into a generator; adapt a fishing box for use as a seat.*
• Recognise the social, moral and environmental effects of technology.	*Considering the effects of a new motorway, intensive rearing, space shuttles.*
• Recognise and take into account in their designing that people can be an element in a system.	

Standard Attainment Tasks

So far we have described all the key elements of National Curriculum Design and Technology except one, the Standard Attainment Tasks; we have noted that this discussion takes place in Chapter 4. But it is important to mention here that each level of Attainment in each Attainment Target has in theory to be matched to a Standard Attainment Task (SAT). However it has been decided not to use SATs for Key Stages 1 and 2 in Technology and so, despite some preliminary work, these have been abandoned. Of course, this does not alter the requirements of specific levels in each Key Stage and teachers will still be required to assess achievement along the lines specified by the Schools Examination and Assessment Council (SEAC).

Work is actively proceeding in the development of SATs for Key Stage 3 as described in Chapter 4. The development of Assessment of Key Stage 4 is linked with the debate about the overall assessment of teaching in GCSE – where many issues remain yet to be resolved. This is discussed in Circular 3/90 as follows:

> *During the Summer of 1995* the first cohort of pupils at the end of Key Stage 4 will be assessed. The main assessment instrument for the National Curriculum at this stage is the GCSE. Many pupils will take courses leading to a GCSE in technology. However the attainment targets and programmes of study for technology are flexible enough to allow a range of other options at Key Stage 4. There will be scope for combined courses in technology and one or more other subjects leading to a GCSE; and for courses in technology alone which take substantially less time than is required for a

GCSE course. There will also continue to be scope for qualifications validated by the vocational examining bodies. The NCC is considering possibilities for combining technology with other subjects inside and outside the National Curriculum for GCSE purposes, and will be issuing guidance in due course.

Pupils will not be allowed to drop technology before the end of the key stage. The Secretary of State has asked the National Curriculum Council to consider what curricular options should be available for children who obtain a good GCSE in technology before the end of year 11. He has also asked the Council to publish general advice to schools on planning the curriculum at Key Stage 4, with examples of good practice.

However, it is important to note that the assessment criteria for Technology at Key Stage 4 are not the same as for GCSE. There will be three types of qualifications in the technology area: Design and Technology (ATs 1–4), Technology, which will involve using IT in designing and making and will emphasise control (ATs 1–5) and Information Systems (AT5 plus additional material). There will also be short courses within these headings. It is planned that exam syllabuses will start in September 1993, the first certification being in 1995.

The growing importance of vocational qualifications at Key Stage 4 – particularly those of a technological nature – has been emphasised by the White Paper on Education and Training (1991). It was foreshadowed by the Secretary of State for Education in January 1990, speaking at the North of England Education Conference.

> Combined courses leading to GCSE are not the only available option. A number of schools offer to

pre-16 pupils qualifications validated by vocational bodies such as the Business and Technician Education Council, City & Guilds of London Institute, or Royal Society of Arts. There has been concern that such qualifications will not be permitted in future.

Any approved qualification which purports to cover a National Curriculum subject in any sense or to any extent will have to cover adequately the targets and programmes concerned. I see no reason, however, why the vocational examining bodies should not be able to submit such qualifications for approval. There seem likely to be particular possibilities for them to do so in technology.

For the foreseeable future, it is clear that the less formal but mandatory teacher assessments of Design and Technology will constitute the main assessment at Key Stages 1 and 2 and will play an important role in Key Stages 3 and 4. However, the general issues of assessment will be discussed fully in Chapter 4.

There is much else in the provisions for the National Curriculum that affects Design and Technology. In particular it is required to overlap and interrelate with all other foundation subjects and in particular with the core subjects of Science, Maths and English, as Fig. 3.1 shows. Links with Science are particularly interesting, as Mansell (1990) notes:

> Comparing the statement of attainment for technology and core subjects provides some interesting possibilities for teamwork.
>
> In AT3 Planning and Making, the importance attached to pupils' ability to make quality products with progressive understanding of the principles involved provides strong links to science, e.g. Design and Technology AT3 2(b), 3(b), 4(c), 5(b), 5(c), 6(d), 7(b), 8(b), 9(b) and 10(b).
>
> The linking attainment targets in science are AT6 – types and uses of materials; AT7 – making new materials; AT8 – explaining how materials behave; AT10 – forces; AT11 – electricity and magnetism; AT12 – scientific aspects of IT, including microelectronics; AT13 – energy; AT15 – using light and electro-magnetic radiation and AT17 – the nature of science.
>
> Cross-referencing the different subject ATs provides one framework for planning schemes of

work, assessing and coordinating pupils' progress. It also helps to fulfil the requirement that 'pupils should be taught to draw on their own knowledge and skills in other subjects'. Developing a coordinated approach will take time, but it will help teachers to see how pupils' learning can be drawn together and provides a basis for continuity and progression.

It would be equally possible to provide a list of links between Design and Technology and all other foundation subjects. For example the capability to provide coherent technical instructions for end users must relate as closely to Attainment Targets in English as to those in Design and Technology. (Any reader who has tried to follow the user's manual for many makes of personal computer or video cassette recorder will be well aware of the importance of this capability!) Medway (1990) writes as an English specialist on the consequences for English teachers that spring from the inclusion of Technology as a National Curriculum subject and outlines the range of new opportunities this gives for the teaching of English.

A wide range of subject organisations and public and commercial bodies have produced documents that can help teachers to develop links across the curriculum. Typical are the documents from the Arts Council Photography Working Party (1990) and the Design and Technology Association (Eggleston, 1990).

There are also many other factors to consider. They include the cross-curricular themes, the cross-curricular dimensions and the cross-curricular skills (featured in Fig. 3.1) and the links with Religious Education noted in Chapter 1 – all these and a multitude of other National Curriculum considerations can and must influence the teaching of Design and Technology.

There are also the important linkages that have to be made within the subject itself. Throughout each Key Stage, pupils' work must relate to a range of contexts, home, school, leisure, community, business and industry and lead to a range of outcomes (artefacts, systems and environments) and provide experience in a range of materials – wood, plastic, clay, graphics materials, food and fabrics, etc.

Support and guidance

There is also a growing amount of support and guidance for Design and Technology and IT. The former Chairman and Chief Executive of NCC, Duncan Graham, noted that this is 'written and edited primarily by teachers concerned with technology education' and addresses the 'priorities set by teachers' and advisers' groups, and NCC committees, and reflects issues raised during the consultation process'. Norman (1990) comments that:

> For design and technology, this meant clarifying the nature of the subject, establishing strategies for managing design and technology, and constructing a scheme of work. These are clearly the central issues immediately facing teachers. Very useful advice is given on, for example, the role of the design and technology coordinator, and possible models for developing design and technology in secondary schools which avoid the 'circus' approach. There are also examples of the development of schemes of work and modules for Key Stage 3 showing outline plans for work in each of the contexts.

Teaching organisation

Some of the most interesting examples of guidance concerns models of teaching. Four National Curriculum Council models are shown in Figs

3.2–3.5 ranging from the traditional pattern of one teacher working with one class through a range of media, module and topic organisations that are more likely to be able to offer a rich and integrated Design and Technology experience. The National Curriculum Council has also outlined the possible delivery of Key Stage 3 (Fig. 3.6). Together these models, like most other examples of non-statutory guidance, indicate the vast range of flexibility that is available within what, at first sight, may seem a prescriptive and restrictive set of legal requirements.

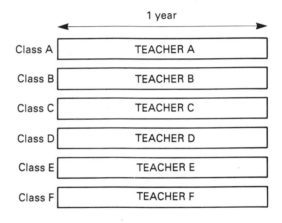

Fig. 3.2   Model A

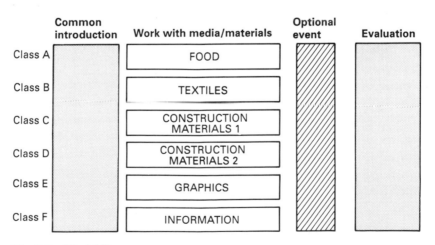

Fig. 3.3   Model B

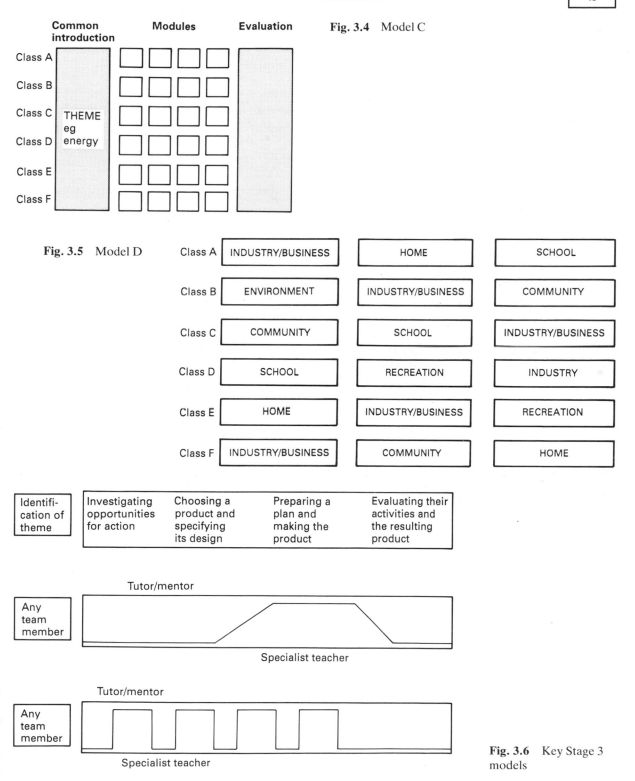

Fig. 3.4   Model C

Fig. 3.5   Model D

Fig. 3.6   Key Stage 3 models

Conclusion

The account of the National Curriculum derived from official documents portrays a firm, assured regulatory stance on all aspects of Design and Technology. Yet as this chapter has intimated, this is far from the case. The actual experience of Design and Technology which each child will receive will be determined by a range of variables. These variables include the availability and use of resources, the priority given to Design and Technology by the school and the enthusiasm and support of the parents and employers. If there is not an all-round desire to achieve the Attainment Targets in Design and Technology fully and other priorities such as achievement in other subjects and activities are preferred, then schools could still deliver pupils at Key Stage 2 with only Level 2 or even at Key Stage 4 with only Level 4 achieved and yet still remain within the law.

But the greatest variable of all springs from the teachers. Their experience, their training – both initial and in-service, their capacity to work in a team to deliver Design and Technology, their management skills and above all their enthusiasm, energy and commitment, are likely to have a far greater influence than any legislation. Indeed, keeping within the letter of the law will not be too difficult for teachers as familiarity with National Curriculum requirements grows. Teaching within the spirit of the National Curriculum is an altogether more demanding task. The National Curriculum points the way to an impressive delivery of Design and Technology but no legislation, however detailed, can ensure it; it can only be delivered by the will of the teachers.

We have reviewed some of the main provisions for National Curriculum Design and Technology. It is not intended as a substitute for the official regulations and non-statutory guidelines but an accompaniment to them. Above all, it is hoped that this chapter has emphasised the continuing role that individual teachers will have, not only in interpreting the requirements, but also in providing the range of essential contexts that surround them.

References

Arts Council Photography and the National Curriculum Working Party (1990) *Photography in the National Curriculum*. London: Arts Council.

Department of Education and Science (1988) *Design and Technology for Ages 5–16* (Interim Report of the Design and Technology Working Group). London: DES.

Department of Education and Science (1989) *Design and Technology for Ages 5–16* (Final Report of the Working Group on Design and Technology). London: HMSO.

Department of Education and Science (1990) *Technology in the National Curriculum*. London: HMSO.

Department of Education and Science (1990) *Circular 3/90*. London: DES.

Eggleston, J. (1990) *Delivering the Technology Curriculum*. Stoke-on-Trent: Trentham Books/DATA.

Her Majesty's Inspectorate (1991) *National Curriculum and Special Needs*. London: DES.

Mansell, P. (1990) 'Linking design and technology and science'. *Design and Technology Times*. Salford: The University.

Medway, P. (1990) *Technology Education and its Bearing on English*. Leeds: Leeds University School of Education.

National Curriculum Council (1989) *Consultancy Report: Technology*. York: NCC.

Norman, E. *et al.* (1990) *Teaching Design and Technology 5–16*. York: Longman.

White Paper (1991) *Education and Training for the 21st Century* (2 volumes). London: HMSO.

CHAPTER 4

Assessing Design and Technology

This chapter reviews the long route between crude normative assessments of Design and Technology to the sophisticated criterion referenced approaches developed alongside the National Curriculum that are available for use if the professional and political will to use them exists.

Throughout the 1970s and 1980s the great spurt in the development of Design and Technology led to dramatic changes in the content of the subject and the way in which it was taught. Yet relatively little attention was paid to the ways in which the new subjects could be assessed. What assessment existed was largely confined to school-leaving examinations – still *normative* in nature, offering only rank orders of successful candidates and only incompletely covering what was being taught and learned. As with all normative examinations they offered only banded scores (A, B, C, D, etc.) but virtually no information on the content of the material being examined or the understandings achieved. Yet it became increasingly urgent for teachers to be able to identify and report the achievements taking place in the new approaches to ensure that parents, employers and the pupils themselves could recognise and value the new learnings and understandings. But it was also important that teachers could diagnose and build upon what was being achieved. The problem was exemplified in a research study of a primary school conducted by Sharp and Green (1975) who demonstrate the need for teachers to have and use a 'Reporting Language'. Although the subject under consideration is mathematics the example could easily be applied to Design and Technology.

The authors report:

Mrs Carpenter had recently switched to 'new mathematics' after teaching a more traditional maths syllabus.

TEACHER: When you've got a set plan . . . everything in its place . . . you taught length immediately after you taught so and so, and it was taught, you know, it was not a matter of children learning really, not in the way we'd been thinking that they should be learning . . .

INTERVIEWER: How do you mean?

TEACHER: I mean we all, well, I have a little plan but I don't really . . . I just sort of, mmm, try and work out what stages each child is at and take it from there.

INTERVIEWER: How do you do this? How does one notice what stage a child is at?

TEACHER: Oh we don't really know, you can only say the stage he isn't at really, because you know when a child doesn't know but you don't really know when he knows. Do you see what I mean? You can usually tell when they don't know [*long pause*]. [*There was a distraction in the interview at this point.*] What was I talking about?

INTERVIEWER: Certain stages, knowing when they know . . .

TEACHER: . . . and when they don't know. But even so, you still don't know when they really don't [*pause*] you can't really say they don't know, can you? . . . . That's why really the plan they wanted wouldn't have worked. I wouldn't have been able to stick to it, because you just don't . . . you know when they don't know, you don't know when they know.

Fortunately the opportunities to develop such a reporting language have been facilitated by the development of *Criterion Referenced Assessment* (CRA) in which assessment is based on pupils' achievement in specified aspects of the subject which can then be recorded. The contrast between Criterion Referenced Assessment and *Norm Referenced Assessment* (NRA) is best explained by an analogy with athletics. CRA indicates the times achieved by pupils in a race and also adds other aspects of their performance such as style so that comparisons can be made to past performance and averages derived for their age and condition. NRA simply records which pupils were first, second and third – and last – and offers no other information.

The Goldsmiths' Project

The opportunities to significantly develop CRA in Design and Technology became available when the DES Assessment of Performance Unit conducted a major exercise to explore how achievements in Design and Technology might be identified and measured. The team, based at Goldsmiths' College under the direction of Richard Kimbell, began work in 1986 within the Assessment of Performance Unit of the DES and from April 1989 as a team within the Schools Examination and Assessment Council. In November 1988, approximately 10 000 pupils were involved in the first ever national survey of Design and Technology capability. Kimbell (1990) writes about the survey as follows:

> A number of survey techniques were used, including GCSE project profiling, half-day design and development 'modelling' activities, and shorter structured pencil/paper design and development tasks. The pencil and paper test activities were supervised by teachers in their own schools. The more complex modelling activities and project

profiling were run by teachers trained by the research team. Teachers from all over the country took part in the assessment and marking of pupils' responses to the survey activities.

Some questions that the research is helping to answer are:

- What do we mean by capability in Design and Technology?
- What are the constituents of Design and Technology capability?
- How can we help to develop Design and Technology capability?
- How can we measure Design and Technology capability?
- How does curriculum experience affect Design and Technology capability?
- How does a pupil's general ability affect their Design and Technology capability?
- How does gender affect Design and Technology capability?

Consequently in the survey we developed test activities that made use of several forms and time scales including:

- Starting with an evaluation activity on two similar products, and moving on to redesign the products to improve their performance.
- Starting with a half developed idea that needs to be developed and detailed.
- Starting with a broad context that needs to be explored to identify starting points for the design team.

The responses from the survey suggest that each of these forms has particular strengths and weaknesses. Taken as a whole the greater the diversity of activities offered by the teacher, the richer will be the experience for pupils.

It is clear that a number of factors crucially affect pupils' performance:

- The context of the task it must be established in a way that gives meaning to pupils, and they should have the chance of working from a variety of such contexts.
- 'Ownership' of the task – pupils must see it as genuinely their own.
- The specificity of the task – if it is too tightly defined by the teacher there is less chance that the pupils will take ownership of it.
- Response to the task – pupils should be encouraged to respond to the task in a variety of

APU Model

The survey has shown that there are two distinct aspects of capability in Design and Technology that need to work together as pupils tackle a task. The two aspects of capability are. . .

Reflective   and   **Active**

(Thinking around the task)   (Taking action on the task)
Seeing and reflecting on   Taking action to develop
the issues that bear on a   proposals for the new
task   artefacts, systems or en-
                                           vironments

The two aspects of capability are linked in an iterative process 'to-ing and fro-ing' between thought and action.

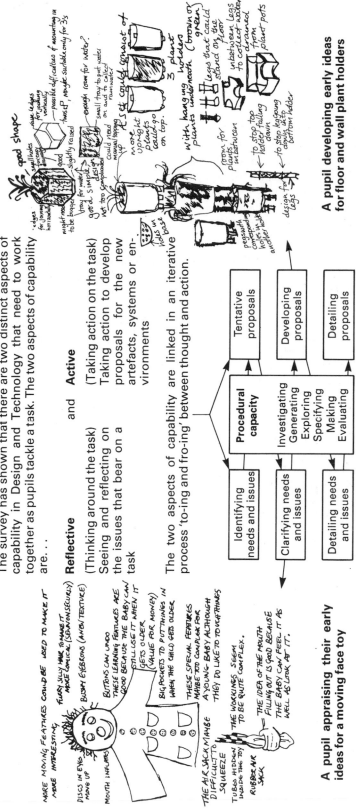

A pupil developing early ideas
for floor and wall plant holders

A pupil appraising their early
ideas for a moving face toy

Identifying needs and issues	→	**Procedural capacity** Investigating Generating Exploring Specifying Making Evaluating	→	Tentative proposals
Clarifying needs and issues				Developing proposals
Detailing needs and issues				Detailing proposals

As this developmental process continues
. . . the task becomes clearer   . . . and the constraints on
                                    the solution become more
                                    critical

ways, e.g. through written, graphic, oral and practical modes.

● Collaboration and Groupwork – pupils should be given the opportunity to share their ideas as well as to work individually.

It soon became clear that a great deal of the work of the Goldsmiths' Project was to do with understanding the nature of Design and Technology, of process such as imaging, modelling and manufacture, the 'holism' of the work and, of particular importance, the motivation of pupils. Major sections of the Final Report (SEAC, 1991) analyse and discuss capability covering such topics as conceptual understanding, representational capability and communication skills.

A typical example of the work of the project can be seen in this extract from the Final Report:

> The test booklets effectively structured the 90 minutes of activity into 5, 10, 15, or 20 minute blocks of activity, and typically there were nine or ten interventions by the administrator during the period, each directing pupils' attention to considering specific aspects of their task. Our reason for doing this derived from our need to generate evidence for assessment purposes (discussed in detail in Sections 7 and 8) but we soon realised that we had developed a system that not only allowed us to assess performance – but also promoted it.
>
> To take a specific example, in one of our tests, pupils are given a restricted amount of space in a booklet and asked to jot down (in 20 minutes) some initial thoughts about possible ways to tackle a task (e.g. the design of a bird scaring device that is powered by the wind and based on the familiar spinning advertising boards). After this 'early ideas' phase, they are asked to list (in 5 minutes) the things that the product would have to do if it is to be successful. They are then asked to look at their early ideas in relation to these issues and (in 5 minutes) jot down what they think are the strengths and weaknesses of their early ideas.
>
> This closely structured thirty minutes of intensive activity stands in marked contrast to typical classroom/studio practice, in which pupils often have long periods of time (40, 50 or 60 minutes) and often it will be unstructured 'design development' time.
>
> Developing and realising a design is a massively

complex activity. One has to keep many issues in mind all the time, gather information from many sources, explore ideas, model them and test them out against the reality of the user's needs. The development must result in an outcome that is achievable in the time available, with the physical resources at one's disposal, and using skills and knowledge that are either directly held or readily acquired. Many pupils find such mind-bending demands to be very daunting, and a major part of the complexity is procedural; put simply 'What shall I do next?'

> In an unstructured activity this complexity is maximised; in our booklets it was minimised. They took the lead in guiding the response of pupils – indicating what sorts of activity were appropriate for this 5 minutes, and how it should be different for the next 10 minutes. The booklets projected a particular repertoire of appropriate responses – not in substance terms of *what* to design – but in *procedural* terms of what sorts of things need to be thought about/done this moment.

The Goldsmiths' team approach may be typified in this comment on making assessments:

> The assessment of pupils' work in the survey began with a consideration of 'holistic' capability, an overview judgement of quality. It is important to remember that teachers were not marking their own pupils' work. Rather we collected all the work from the survey and distributed it to teams of markers who were quite unable to identify the pupils concerned. Even the pupil gender was not known, for pupils were represented as just a number.
>
> Given this situation, there were many who expected the reliability of these overview 'holistic' judgements to be low. We felt however that there was a well established tradition in design and technology of holistic *teaching* (on whole projects from task clarification to evaluation of end product) and that this ought to enable teachers to hold an overview of what capability is like.
>
> However, it is one thing to establish a position of principle, and quite another to show that it can lead to reliable assessments. When 200 pieces of work had been marked (on a 6 point scale) by every marker, we jumbled them all up again and sent large samples to a second batch of markers. 'Correlations' are a measure of the match of the two rank orders. A correlation of +1 represents a

perfect match and −1 a complete reverse or mis-match. A zero correlation is simply a random distribution. The details of the correlation analyses can be summarised by reporting that the median correlation was above 0.7, which – given the nature of the judgements in question – is a very respectable result. It suggests that given appropriate training it is possible to pick out and agree on a measure of consolidated holistic capability.

It will be seen that the Goldsmiths' team has, to quote their own conclusions, found it impossible to 'constrain our activities to the single area of assessment'. This breadth and the inevitable lack of close articulation with the National Curriculum Design and Technology framework (which was only established in 1990) makes it difficult for teachers to use the Goldsmiths' report in any direct way to assist their role as assessors. This is in no way to devalue the project's work indeed the new understanding of teaching strategies developed by the group and made known through its leaflets (*Negotiating Tasks, Structuring Activities*, etc.) constitutes a major contribution to the subject and will probably be the most enduring testimony to the research work.

The Goldsmiths' holistic approaches are, however, being modified and to a considerable extent being applied in the development of Standard Attainment Tasks for Key Stage 3 and Teachers' Assessments (TA) (SATs for Key Stages 1 and 2, Technology having been discontinued and work for Key Stage 4 awaiting clarification of GCSE issues; these will be discussed later in this chapter).

Assessment of National Curriculum Design and Technology

For the immediate roots of assessment of National Curriculum Design and Technology we must turn to Paul Black and the TGAT (Task Group on Assessment and Testing) Report (1989). It is here that the guidelines for the National Curriculum Assessment were laid down and, though since modified, it is still the baseline from which SATs are being developed. Black notes three main ideas from the report:

The first of these was to stress the formative aspect of assessment. This was part of a view that assessment must be an integral and helpful part of teaching and learning. Many teachers say that they welcome the new curriculum but are worried about the assessment. Whilst such worries are understandable, they often go along with a view that the assessment is an extra imposition, which good teaching and learning could do without. This is quite alien to the TGAT philosophy. Problems about whether there will be too much assessment in the new system will, in the long run, be answered by teachers themselves as they determine the optimum amount.

This optimum will be achieved when the attention given to checking and recording pupils' progress so helps to meet their learning needs that it justifies the time that it requires.

The second TGAT idea is that good assessment depends upon using a range of methods. Artificial and narrow methods of assessment are undesirable both because they give unreliable results and because they can have undesirable feedback effects on teaching.

The third TGAT idea was that national assessment results should be determined by a combination of teachers' assessments and the results of external SATs. The SAT development teams are developing their methods in close collaboration with schools. This is because they need to ensure that the SATs make sense to pupils, are feasible for teachers to administer, and are consistent with the teaching interpretations of the national curriculum which are emerging in practice. Such an approach is essential, both because of the novelty of some important aspects of the curriculum and because of the lack of a body of experience of appropriate styles of assessment. Thus, as teachers' own assessment practices are developed, they will influence the SAT development to a far greater extent than many anticipated.

The exact way in which teachers' assessments and SAT results will be reconciled is still not clear. The degree of teacher responsibility for the SAT outcomes, and the classroom validity of the SATs, may narrow the number of discrepancies. It will nevertheless continue to be true that teachers' assessments will be an important component and that they may be preferred to SAT results if examination of the evidence can justify this. All of these developments are consistent with the TGAT argument that teachers' own assessments, conducted in familiar contexts and over extended periods of time, have potential reliability which SATs can only match if they are designed to share some of these inherent strengths.

However SATs for Design and Technology present their own problems as Black has noted. The Working Group recommended that Technology be based on pupils tackling projects as a whole, allowing achievement in a variety of tasks – rather than in separable activities. This gave rise to a further variable – projects which emphasise originality and inventiveness do not have 'correct' solutions so that a wide range of accepted responses is not only likely it is also desirable. Black (1990) comments:

> This led the working group to argue that even a bank of standard project tasks could hardly be acceptable, because any specified task must presuppose that the need or opportunity has been defined, so pre-empting the outcome of attainment target one. They did however accept that 'some more narrowly focused SATs' might play a subsidiary role in assessing certain aspects of the subject. Finally, they stressed that the particular lists of knowledge and skills set out in the programmes of study were not to be assessed directly. However, the proper use of appropriate knowledge and skills ought to be an important aspect of assessment of any task, for the use of knowledge and skills to a high level is essential to good technology work.

These recommendations leave many problems to be resolved. The easiest way to approach these might be to start with the idea of focused tasks. It is fairly easy to foresee how limited assessment exercises could be constructed to concentrate on particular aspects of problem-solving in design and technology. The following list sets out some (by no means all) possibilities:

- presentation of data about a broad problem area so that pupils could show their ability to work on needs and opportunities
- perhaps in combination with the above, a design study
- presentation of a problem and a proposed solution for pupils to evaluate
- appraisal of the historical and cultural influences on a product

- critical reflection on the strategies and concepts underlying technological developments, perhaps in relationship to particular historical examples.

All of these are feasible and each could reflect an important aspect of the curriculum. The critical question to ask about them is surely whether or not similar activities would be a valid part of good teaching in the subject. I think it is clear that they could be.

However, it is also clear that if the learning of design and technology were to consist exclusively, or even mainly, of such atomised activities, the main spirit and aim of the National Curriculum would be lost.

Assessment at Key Stages 1 and 2

At Goldsmiths' College a team directed by Kay Stables is working on Key Stage 1 non-statutory SATs and TA guidelines. (There are no proposals for statutory SATs at Key Stages 1 and 2.) Material developed by the Goldsmiths' teams has been made available to all schools including tasks such as protective headgear, plant and animal shelters and 'setting-up shop'. Welsh versions have gone to Welsh schools. Detailed working for Key Stage 2 starts in 1992–3.

Assessment at Key Stage 3

Following the leads of the TGAT Report and the considerations of the Schools Examination and Assessment Council (SEAC) two teams were chosen to develop Standard Attainment Tasks for Key Stage 3.

Both teams were parts of major consortia. One was the London-based Consortium for Assessment and Testing in Schools (CATS) team. Here the work in Technology was being undertaken at Goldsmiths' College directed by Richard Kimbell and Jim Patterson, previously of the APU team, working with a newly constituted group of researchers. The Midlands Examinations Group National Assessment Project (MEGNAP) Key Stage 3 team was based at Middlesex University under the direction of Richard Tufnell and John Cave.

The SATs developed by the teams will not of

course be published and, in their 1991 version, may not even be used, but some unpublished comments by the Middlesex University team on their approach and their pilot provides a useful guide to the problems encountered, for example:

Timetable allowance

Comments highlight our concern about what pupils have actually learnt in relation to the skills, knowledge and procedures which are generic to Design and Technology. The lack of evidence in pupil outcomes of these attributes could be due to a lack of curriculum time – there is certainly a wide disparity in timetable provision. In each year of the Key Stage there is evidence of the time allowance varying between 24 and 72 hours for both of the main providers – HE [Home Economics] and CDT [Craft, Design and Technology]. Consequently some pupils in the trial may have had, in broad terms, three times the experience of other pupils throughout the Key Stage. The variation in time allocation is similar for Art but in many schools it is not being incorporated, as yet, into the Design and Technology pool. There is some evidence that there is a timetable provision for Information Technology but it is spasmodic and rarely amounts to more than 20 hours a year. There was no provision for Business Studies in Key Stage 3 in any of the schools in which we trialled.

Resources

The resources available for Design and Technology must also be a key factor. In the '91 pilot we wish to investigate this issue in a more focused way – does capitation allowance per pupil affect performance? As with all aspects of this subject the variations are spectacular. We asked teachers to tell us how much was allowed per pupil in Y9 for consumable materials for their subject. The range was 40p to £6, so wide that there must be some effect on pupil options and possibly performance. Perhaps the best overall picture can be gained from the four cohort schools where all Design and Technology teachers completed questionnaires. In these schools the range was between 40p and £2.50 with an average of £1.22. It is difficult to imagine how teachers provide about 38 sessions (average time 68 mins) in what is still considered a practical element of the curriculum on amounts as meagre as these.

The picture illustrated by these findings leads us to conclude that the SAT experience which we operated was a novel one for both teachers and pupils. It relied mainly on CDT and HE teachers and the facilities in which they normally operated. It is our perception that it is teachers from these two subjects who will be the prime providers of Design and Technology in Key Stage 3. We would draw the conclusion that curriculum INSET [In-service Training] should focus on providing teachers from these subjects with the skills, knowledge and flexibility required to deliver NC [National Curriculum] Design and Technology. The resources and facilities available to pupils were inconsistent and varied and we believe must be borne in mind constantly when dealing with issues such as reliability.

SAT evidence

Our concern about the effect of resources on SAT outcomes is in fact part of a much wider concern about what pupils actually do during a SAT to demonstrate capability in Design and Technology. Pupils' work to date, although encouraging in some aspects, shows very little evidence of those specialised characteristics that are distinctive of, or generic to, Design and Technology. This shows up most clearly in the obvious lack of ability of pupils to communicate through drawing. We have found only one piece of evidence, for example, that a pupil has knowledge of a specialist (technical) drawing convention – although many generally more able pupils are able, for example, to write copiously about what they intend and have accomplished. (This is all the more worrying since most pupils have experienced three years of HE and CDT which ought to have made a more significant impression on SAT outcomes.)

Throughout the development phase we have been attempting to make it possible for teachers to take ownership of the SATs by careful design of the material itself and through INSET – culminating in the INSET video that features interaction between teachers and pupils in a variety of different schools. Our message is clear that throughout a SAT teachers should be teaching but must avoid pre-empting those decisions and actions that will provide evidence of individual capability. In earlier trialling, some staff treated the SAT as a closed examination but we hope this problem has now been eliminated.

Staffing and organisation

Generally, it seems that specialist staff have

cepted the principle of inter-subject co-operation to facilitate a comprehensive delivery of Design and Technology. In practice, different 'experiments' with curriculum re-structuring have brought mixed results, but given the number of variables and our sample size, this is perhaps not surprising. We have clearly established that CDT and HE are the two areas principally involved in delivery (Section 1) and specialists within these see their future firmly within the subject confederation at the level of Key Stage 3. Art and Design specialists often appear less certain about their role and many are clearly looking towards the emergence of their own order for the subject. We have found very little evidence of teachers with a Business Studies interest involved in SAT delivery, and in a few instances teachers have been drawn in who have no relevant subject knowledge. It is very important to note that where subject areas have come together for SAT delivery, the resulting mix of teachers by no means guarantees pupils equal access to the specialists they may require.

Both SAT teams developed sophisticated strategies to devise fair, searching and reliable tasks for assessment. To achieve this they developed 'long' SATs involving a minimum of six weeks' work by pupils.

The techniques adopted were complex. For example, the Middlesex team experimented with stranded matrices in which the assessment model was produced by mapping across the Statements of Attainment from the Statutory Orders to develop a model for trialling which revealed distinct areas of technological competency. An assessment matrix was then produced with these strands arranged as vertical columns and the levels of attainment as the horizontal bands. It was hoped that this matrix would prove helpful as a diagnostic tool. The identified strands of competency according to Attainment Target were:

AT1 Identifying possibilities
     Carrying out investigation
     Recognising implications

AT2 Developing designs
     Decision making
     Communicating and modelling ideas

AT3 Competency with materials
     Manufacturing capability
     Organisation and planning

AT4 Evaluating of technological activity
     Appreciating and appraising technology

In each strand of competency at each level there is either a single statement or a group of statements in a box. Some of these boxes spread across more than one strand. These stranded matrices were contrasted with other matrices based on the Attainment Targets and the Statements of Attainment directly.

However by mid-1991 it became clear to both teams that the sophisticated, complex SATs they were developing were unlikely to be acceptable. Following the first full application of SATs for Key Stage 1 English, Mathematics and Science there was a widespread professional demand for shorter, less time-consuming SATs. This was welcomed by Ministers who were concurrently indicating a preference for short, predominantly paper-and-pencil tests.

The matter was discussed at length by the MEGNAP team as one of their internal documents (quoted with permission) makes clear:

It is quite apparent that a SAT in the form which we have piloted this year will not be acceptable in the future. SEAC has established a set of criteria (from the Secretary of State's speech) which they believe a SAT should meet. The framework indicates that a terminal examination should be:

simple – rigorous – objective – time limited
manageable for both teaching programmes
capable of being administered to pupils
   simultaneously

Although the Secretary of State also referred to the possible inclusion of a practical, or project work element SEAC do not appear to favour this approach. Our overriding impression is that a SAT is required which will fit into a one and a half hour time slot – an end of term exam. They are currently investigating the notion that all subjects should try to fit this pattern and only in exceptional circumstances will an alternative be acceptable. Any new specification, we think, will revolve around a very similar specification for all subjects.
  We obviously have to review what we should do

in the light of these possible changes. If we are involved in the next academic year it is certain that we will be working to a new specification. We have some opportunity, however limited, to influence that specification. It is important that we start to review our developments and determine ways in which we could try to do what is being asked.

The rationale is simple. In the past children studied a syllabus. An examination then sampled, at random, the content. It was then assumed that performance in the sample was a sound indication of overall competence. The question is being asked, why should National Curriculum assessment be any different? Our initial response would be that we are working to TGAT – which changed the nature of assessment. Ministers now respond that TGAT was the starting point but not the solution. Management of assessment is a key issue and TGAT failed to appreciate the scale of its proposals.

We would also respond that Design and Technology is different (you know the arguments, I do not need to rehearse them). The most powerful difference, which SEAC might recognise, is the novelty of the subject. Our SATs had to create Design and Technology before we could assess it – SEAC and DES would accept that point of view. However they would respond that by 1993/94 this will not be the case. By then schools will be delivering NC Design and Technology so there will be no need for SATs which provide curriculum INSET – which is how SEAC might describe our SAT. They might also respond by saying that if TA [Teacher Assessment] is based on whole activities why replicate the whole process again? Why indeed, providing you can assure the quality of TA and the basis for those assessments?

Traditionally examinations produce a rank order – not an item of assessment information which the NC attempts to promote. The NC is concerned with telling pupils and parents about performance in relation to national standards. Teachers are very good at producing rank orders on the basis of comparative assessments and an exam might capitalise on this.

What questions need to be asked of this new style of SAT?
Will it be possible to assess all ATs [Attainment Targets] and SoAs [Statements of Attainment]?
Will it be necessary to assess all ATs and SoAs?

Will an exam of this nature, by necessity, be primarily reflective and written?
Will it be possible to assess all ten levels via an exam?
Will the style of the SAT influence the nature of key stage delivery?
Will pupils with special educational needs be able to cope with exams?
Will an exam be able to reflect the breadth of Design and Technology, does it need to? etc.

Any exam of this nature is going to provide a quality control procedure of teacher assessment – a validation procedure. The SAT will be a policing tool, not something which exemplifies and promotes the subject. It is very difficult to imagine how a SAT of this character could produce holistic Design and Technology assessments but this would not be its objective. A dipstick approach will sample capability and make assumptions!

What are the possible models?
The team identifies three, based on reporting, reorganising and reacting and reflecting specifically:

1 A curriculum activity, not a SAT, followed by an exam, which constitutes the SAT. The curriculum activity might consist of a SAT similar to those piloted this year. It would be described as an activity/task framework to which schools must adhere. Schools would undertake the task when it suited them, during Y9, but obviously prior to the exam date. The framework would characterise sound Design and Technology practice. Pupils would then use the outcome as the starting point for the exam. Questions in the exam would attempt to assess a bundle or string of SoAs from an AT which describes progression in a particular aspect (these strings may consist of one or two statements or as many as four or five). In an exam of this type it must be possible to ask a question which would be applicable to all statements in the string. The quality of the response would determine which level of the statement had been secured. The first step is to establish the number of strings which can be identified and establish the framework for such a testing approach. This style of examination would confirm a range of SoAs which a pupil has secured. For example AT2 4b, 5b, 6b, and 7b:

As part of a task on measurement a height

chart has been identified as being a good idea for helping children understand how they grow. The designer has proposed a printed paper chart. If you were the designer what would you do next?

2 A specifically focussed assessment of one AT. This could be an active task as well as a reflective one. It would, within an activity, look at a pupil's levelness of operation. Any questions would need to be closely related to the over arching AT descriptor. This type of exam would interrogate a pupil's experience and may involve the task of looking at a real or hypothetical 'pathway' through an activity. It would produce a single AT score against which TA could be scaled. For example: *A questionnaire is required to assess people's viewing habits. Generate a design specification, explore ideas, produce a proposal and develop it into a realistic, appropriate and achievable design.*

3 An exam which reviews all work in the key stage and attempts, by a series of questions, to validate TA by asking pupils to transfer experiences into a new context.

This approach might exploit strands as we have identified them. Level would be determined by outcome. A range of questions would examine the breadth of a pupil's experience. Questions in this framework would attempt to address all ten levels.

This style of examination would also review the process of Design and Technology and would cover all ATs. It would focus on a pupil's theoretical understanding of how to approach any task. Coursework would provide a practical backdrop from which illustrative examples could be drawn. This exam might produce a guide to levelness in each AT. For example, synoptic approach: *Why is planning an important factor when deciding how to satisfy a need or opportunity you have identified?*

Despite or perhaps because of such considerations, however, there has been a further change of direction by Government to approve, after all, wide-ranging Design and Technology SATs, providing that an acceptable examinations component is included. The official announcement in January 1992 specified a 1½ hour written test supplemented by a statutory practical test of manufac-

turing skills lasting between 10 and 12 hours and conducted during ordinary lessons. The Middlesex team is again developing these, but alone; the Goldsmiths' team is now concentrating on their Key Stage 3 SAT contract. For their second contract the Middlesex team is also working independently and not as part of the MEGNAP group.

Teacher Assessment

Although Design and Technology Key Stage 3 SATs from 1994 onwards are still unpredictable the requirement for Teacher Assessment (TA) is binding on all teachers at all key stages. There is no doubt that the detailed work of the two teams will be of immense assistance in this work. There are already interesting beginnings of Teacher Assessment – particularly in the early years as Wendy Pitt reports (1990):

Many opportunities to develop Design and Technological activities arise from the thematic work taking place in the classroom.

Such an opportunity arose when a class of six year olds were working on the theme of *The Teddy Bears' Picnic*. The children had made paper Teddy Bears, coloured them in a variety of ways and produced a class frieze and read and written stories about Teddy bears. But now the experiences of the children were to be broadened.

They had made paper Teddy bears, so the use of a different material was the next planned learning experience. How could we make a display of card Teddy bears that moved in some way? This was the planned challenge that would enable the children to design and make a card Teddy, building on their experiences with the paper Teddies. The materials were gathered together centrally before the children came in. Also prepared were some handy hints for the children to refer to when they were working on their Teddy bears.

The challenge was introduced to the whole class by reading part of a story about a Teddy bear. It was planned that after they had been given the challenge they would then get into small groups to carry out related tasks and the Teddies would be made throughout the morning.

They were also expected to draw a design on

paper of what they wanted their Teddy to look like before they started making them.

There was a need to show certain groupings of children the hint about the linkage as they wanted to make a Teddy with a moving arm or leg. The children used the plastic Meccano that was available to help them understand ideas about simple linkages.

There was great excitement as each group succeeded and a Teddy with a moving arm or leg appeared. One grouping of two emerged with all four limbs working from only one lever. One grouping produced a Teddy that was almost professional in its finish and appearance. Finally all the groupings had made a Teddy that moved in some way. These were displayed around the class so that everyone could see how the others had made their Teddy.

AT1 Identifying needs and opportunities

There was little that I could assess of children identifying needs and opportunities within this activity. The needs and opportunities were identified by me. I planned this activity to enable the children to develop new skills and techniques and to use tools and materials new to them. There would have been other opportunities within this context to enable the children to identify needs. What do the children like about their own Teddy bears? What food are the Teddies going to take to their picnic? How can they get the food to the woods? What games could they play?

AT2 Generating a design

I asked the children to make a plan of their Teddy before they started making it. In my experience, when children are young, inexperienced or immature, the pencil and paper 'planning' comes during and at the end of the activity. As they become more experienced they can plan at an earlier stage. They often put their rough sketches into the bin 'Because they are on scrap paper'. I need to encourage children to develop portfolios for their designs and technological activities. We could have looked at the children's own Teddies and pictures in books. One of the groups produced a Teddy where all four limbs moved. The difficulty here was to discover 'Whose idea was it?' because whoever it was was operating at Level 6c! The only way I could find out was by talking to the group concerned.

AT3 Planning and making

Outcomes that I can see, photograph and touch are easier to assess than other kinds of outcome although there is still the difficulty of how I assess individuals whilst they are working with others. It would be sad if children had to work individually all the time to allow for assessment to take place. Perhaps I will become more skilled in observing children whilst they are on task so that group work can still take place. How does one assess outcomes that are produced from the same stimulus but are completely different in solution, e.g. children who make a picnic food transporter from cardboard boxes, dowel, cotton reels, elastic bands, etc. and children who make their picnic food transporter from a construction kit?

AT4 Evaluating

Evaluation is something that occurs at all stages during the tasks. They are evaluating the materials to hand, they are talking about why they do things. The opportunities here for assessment come from listening to the children speaking. There are also occasions when assessment can take place when the children have finished their task and questioning by the teacher can help in the assessment of the child. It would have been useful for the children to have brought in their own Teddies and talked about them.

All assessment carried out should be useful and help teachers to plan the learning experiences for the children. But, as I know in my experience, the planned curriculum is frequently not the experiences the children actually receive.

The Schools Examinations and Assessment Council (SEAC, 1990; 1992a,b), have produced an extensive range of materials to help all teachers to achieve effective, valid and reliable assessment. These include, *A Guide to Teacher Assessment, Packs A, B and C, School Assessment Folders* for each key stage and a guide for Local Education Authorities, *National Curriculum Assessment.* These were followed by two particularly helpful publications: *Childrens' Work Assessed: Design and Technology and Information; Technology Key Stage 1* and *Pupils' Work Assessed; Technology Key Stage 3.*

These documents offer highly detailed guidance

to teachers and well illustrated examples of pupils' work. A typical example, from Pack C is:

Forms of evidence

The evidence that a child has achieved a Statement of Attainment may take any of a variety of forms – written, oral, practical – depending on the mode of response appropriate to the statement.

A teacher needs to be satisfied that a child has achieved the level of response specified in the Statement of Attainment; the form of evidence is not itself significant. (See Critical Incidents, 4.5, and Systematic Observation, 4.3.)

Evidence might be available in work books or classroom displays. If children annotate work with information about the task, the books will help to authenticate evidence. Classroom displays are often renewed, and can provide continuing evidence of attainment, particularly in cross-curricular context.

Finding evidence for attainment which is not tangible may cause greater difficulty. Teachers may observe processes which occur. The records they keep of these processes being displayed then become the evidence on which the assessment is based. These records may start with ticks or other symbols, but this is unlikely to be sufficient as evidence for assessment. What is required is fuller annotation, perhaps including:

- a note of what happened (see Critical Incidents, 4.5);
- a note of children's comments (see Sharing Responsibility with the Child, 2.5);
- the teacher's own descriptive comment.

For some Statements of Attainment, several assessments are required before the teacher can be confident that the attainment has been achieved. Some interim recording may be desirable, probably through a schedule of observations. The annotations can be made on the schedules, or in record books. (See Formative Record-Keeping, 7.3, and Summarising Attainment, 8.4.)

This illustrative material needs to be considered in conjunction with the tests produced for TA by SEAC. The first set of such tests for Technology at Key Stage 1 were issued in January 1992 deriving from the Goldsmiths' work. However other tests may be generated by other bodies. Whilst TA at each Key Stage is mandatory the use of these or any other specific tests is optional, and SEAC hopes that teachers will dip into the full range of tests available.

Assessment at Key Stage 4

Assessment of Key Stage 4 at age 16 is still dominated by GCSE and is likely to remain so until the relationship between this examination and the SATs and teacher assessments applicable to Key Stage 4 are clarified. The situation is complicated by the following:

1 There is a very wide range of GCSE programmes springing from the component subjects of Technology and the six GCSE examination boards. Specifically, well-established GCSE examinations exist in Technology, Home Economics, Art & Design and Business Studies. Even in one of the original subjects – CDT – there are three widely available GCSE courses, viz. Design and Realisation, Design and Communication, and Technology. In 1988 these examinations alone attracted 152 675 candidates, half taking Design and Technology, a quarter taking Design and Communication and a quarter taking Technology. All strongly emphasise design; indeed the approach adopted by most syllabuses is very similar to the design process required by the National Curriculum Design and Technology. Yet none wholly cover the four Attainment Targets and the programmes of study for Design and Technology. To adapt and develop the wide range of courses so that they are still delivering a wide range of acceptable as well as negotiable qualifications and also enable the National Curriculum criteria to be satisfied is likely to take a considerable amount of time despite the high priority being attributed by SEAC.

The difficulty is reinforced by the need to integrate the existing GCSE courses in Business Studies, Home Economics and Art and Design. And there is of course an understandable unwillingness to modify well-established and well-regarded examination programmes.

2 There is a belief, subject to a good deal of

questioning, that Key Stage 4 pupils should be allowed to specialise in a range of subjects more limited than the 10 foundation subjects. Whilst the three core subjects, Maths, Science and English are likely to stand, all the remaining seven are unlikely to be mandatory for all pupils. The criteria whereby pupils may drop any of the other non-core subjects will be determined along NCC/SEAC guidelines. So far ministers have indicated that Technology will remain compulsory but may be part of a package of intensive work in a limited range of subjects with a view to high level study at 'A' Levels which may even be taken by some pupils in Key Stage 4. Such pupils may of course exceed Level 10 ATs – in the subjects they take! Yet a further problem is that as they reach Key Stage 4 it may be that it is felt that some children lack the ability to work effectively in all foundation subjects. Here again there may be a case for a reduction in the number of non-core subjects.

3 Yet another problem is that the pupils may be taking other examinations at 16+ such as those of the Business and Technology Education Council. There is considerable enthusiasm that these more vocational examinations be seen as equal status with the more 'academic' GCSE. If a student is taking, say, a BTech course in Business Studies or Technology then they too could achieve a higher level in some aspects of the Technology programmes of study and less in others. This may give rise to skewed Technology courses in which some of the SATs can be over-achieved and others under-achieved.

Overall, it is clear that a long-term process of adjustment is taking place. The 1991 *White Paper on Education and Training* added considerable impetus to the moves for equality of status between academic and vocational examinations in a bid to overcome the traditional divisions which have been described in Chapters 1 and 2. This has led the National Council for Vocational Qualifications to develop new ground rules for broader-based, more general vocational qualifications to be delivered not only by BTech but also the Royal Society of Arts and the City & Guilds Institute. These developments have major significance not only for assessment of Design and Technology but also for its delivery at Key Stage 4 and earlier. This may come not only with the change in examinations but also the further fine tuning of Attainment Targets, Programmes of Study and Standard Attainment Tasks which could well require amending legislation. However, for most children, Technology will continue to become a subject throughout Key Stage 4; regular Ministerial announcements emphasise the importance that is attached to it. And in addition to any examination assessment and any SATs there will always be the need for Teacher Assessment – a mandatory requirement. Fortunately, the Teacher Assessment strategies that have already been developed for examination work in GCSE are well advanced and highly relevant to what is likely to be needed.

Conclusion

This chapter has followed the path of assessment in Design and Technology from simplistic, normative appraisal of conventional curriculum requirements through to the prospect of CRA, SATs and sophisticated strategies of teacher assessment. However many problems remain, not least in the 16-year-old age group where the changing nature and role of the school leaving examinations complicates the scene.

Yet whatever the future developments it is clear that teachers' roles in assessment are crucial. They will not only conduct the vast majority of assessment but will also powerfully influence the ways in which children approach assessment and thereby generate the achievement on which assessment is based. And of course it is they who will determine the use to which assessment is put in subsequent teaching and learning of their pupils.

References

Black, P. (1990) 'National assessment for Design and Technology capability'. *Design and Technology Times*, p. 15. Salford: The University.

H.M. Government (1991) *Education and Training for the 21st Century* (2 volumes). London: HMSO.

Kimbell, R. (1990) 'Learning through Design and Technology'. *Design and Technology Times*, p. 14. Salford: The University.

Pitt, W. (1990) 'Assessment is a picnic?'. *Design and Technology Times*, p. 8. Salford: The University.

Schools Examination and Assessment Council (SEAC) (1989) *Report of the Task Group on Assessment and Testing*. London: SEAC.

Schools Examination and Assessment Council (SEAC) (1990) *A Guide to Teacher Assessment Packs A, B and C, School Assessment Folders, National Curriculum Assessment*. London: SEAC.

Schools Examination and Assessment Council (SEAC) (1991) *The Assessment of Performance in Design and Technology* (Goldsmiths' Project). London: HMSO.

Schools Examination Assessment Council (SEAC) (1992a) *Children's Work Assessed: Design and Technology and Information; Technology Key Stage 1*. London: SEAC.

Schools Examination Assessment Council (SEAC) (1992b) *Pupils' Work Assessed; Technology Key Stage 3*. London: SEAC.

Sharp, R. and Green, A. (1975) *Education and Social Control*. London: Routledge.

CHAPTER 5

Gender, race and Design and Technology

This chapter reviews the problems of making Design and Technology equally and effectively available to all pupils – girls and boys, black and white. It argues that there is no basis for the differentials in availability that persist.

Throughout this book there has been the assumption that achievement in technology generally and in Design and Technology in particular is available to all children up to the fullest extent of their capability. But in many areas of Technology male achievement far outstrips that of female, white achievement exceeds that of black. As there is no evidence that overall ability in Technology is limited by gender or ethnic variables, there is an urgent need to examine the problem – not only on the grounds of justice but also on the grounds of economic necessity in order to satisfy the growing need for technological capability in society.

In principle, the problem should be solved by the very legislation that has brought the National Curriculum into being. For this prescribes that Technology should be taught to all children, black or white, male or female. But this will only be an incomplete solution until we understand the sources of the powerful social pressures that have, for generations, differentiated technological achievement by race and gender.

Gender issues

Let us begin the analysis by considering gender. It is easy to say that Technology is a male subject and that the images of Technology generally portray men doing the physically arduous or intellectually arduous technological activities. Even very young children already perceive the subject in this way. Smithers and Zientek (1991) undertook a national survey of 259 boys and 247 girls, all aged five. They showed, for example, that car repairs and woodwork are seen as almost exclusively the province of men, and mending and washing clothes the domain of women. Both the boys and the girls saw it that way.

But, of course, the popular definition of Technology on which the 'male' image is based is an incomplete one. Food technology, fashion, office skills, are all part of technology and are predominantly undertaken by women. However, even when this is recognised, it is still true that there is much gender stereotyping *within* technology. National Curriculum Technology, including Design and Technology, embracing Home Economics and Business Studies, Craft, Design and Technology and Art and Design is however remarkably free from such stereotyping and the broad definitions of Technology in the National Curriculum is wholly welcome. But by embracing subjects that were previously 'female' areas of the curriculum it may make it all too easy for schools to deliver Technology to boys and girls in separate ways that are little different from when the boys did woodwork and the girls cookery and needlework. The only difference would be that they would now be by 'choice' and not by design. And as most school administrators know, choice can be a very effective strategy for achieving differentiation by apparently democratic means. It is all too easy to build on the perceptions of the five year olds that Smithers and Zientek surveyed. If any reader doubts this then the experience of the author in a TVEI school in South Wales makes the point. There the school decided to make Technology available to all and began its programme with a course in keyboard skills for boys and girls sitting side by side. After this introductory term, the children had a choice of Computer Studies or Office Studies. All the boys chose Computer Studies; all but two of the girls elected for Office Studies. After two weeks into the new course, both girls withdrew from Computer Studies and reverted to Office Studies. The social pressure of the peer group was hugely powerful – and totally successful in confirming gender stereotypes.

There are many other examples. On graduation day at many universities, the solid ranks of male engineering and applied science graduates are scarcely broken by a few isolated women. Their scarcity is emphasised by the particularly loud cheer they enjoy as they mount the rostrum to receive their degrees. The applause is for their deviant gender achievement rather than for their academic prowess. Science like Technology is not a wholly male preserve; women have their place but it is usually restricted to biology. The state of affairs prevails despite the strenuous efforts of university administrators to recruit more women students and of university staff to retain them in technology courses once they are there.

Examples such as these indicate that it is not so much that schools and colleges and universities deliberately reinforce social pressures that the boys go into men's areas and girls into women's areas. Indeed the examples show that they make some attempts to fight these pressures. But the social pressures are so immense. As we have seen, gender expectations of role and occupation are generated from the earliest years of infancy and

are reinforced by parents, communities, employers, other children, the media and society at large (Delamont, 1980). Schools and teachers alone cannot change these pressures but they can make a powerful contribution to doing so. But to achieve this they have to be much more effective than has generally been the case.

An enquiry by the Girls and Technology Education (GATE) Project (1981) showed the extreme differences in gender in entries to GCE 'O'-Level examinations (Fig. 5.1). The work of the GATE project is reviewed by Harding (1988) and the work of the parallel Girls into Science and Technology Project (GIST) is reviewed by Catton (1988). Much research has been undertaken on these issues, typical is Doornekamp (1991) who examines the different ways in which girls and boys come to identify themselves in regard to Technology and the ways in which these are 'self-fulfilling' – so that actual achievement becomes measurably different.

Many researchers demonstrate the reality of gender differences in achievement in Technology but there is slight evidence that the differences are innate and truly determined by fundamental differences in ability or capability. Kimbell (1991) does however suggest some differences in approach to technological problems that are not easily linked to socialisation but these do not relate to differences in basic capability.

The research literature gives rise to a number of teacher strategies that are specifically relevant to the teaching of Design and Technology. These include the following:

1 *A strong emphasis on eliminating gender differences in primary education (Key Stages 1 and 2).* The opportunities for such an approach in primary education – where very few children have studied Technology and few teachers have taught Technology – are great. There is the chance to make a new beginning, uncompromised by entrenched assumptions. This is in marked contrast to secondary education where teachers work in schools which have for long delivered gender-differentiated Technology.

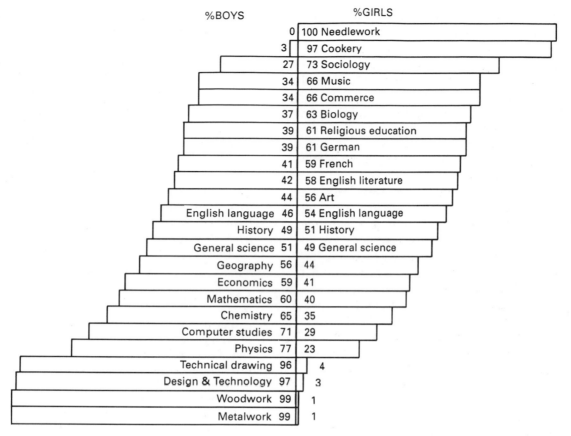

Fig. 5.1 GCE 'O'-Level entries, Summer 1980

Some of the approaches have been described by Brown (1991). She argues that girls need:

- more time on their own initially to gain experience of things which are already familiar to boys
- time to talk over their ideas with their teacher and help to wean them away from the tendency to hang back
- equal access to materials and equipment, particularly those they are not familiar with, such as hard materials like wood and metal. This should be carefully monitored, especially in group work situations
- equal access to help and encouragement from teachers and other adults – time given to boys

and girls should be monitored to ensure equality of attention

- activities which add a scientific and technological dimension to everyday objects or events which are already familiar to them, for instance making a working model of a swing
- to participate in scientific and technological activities which are socially or environmentally relevant such as designing a footbridge to provide a safe crossing over a busy road
- encouragement to recognise that successful achievement in scientific design and technological activities is valued highly like good writing or mathematical problem-solving

To help girls realise their potential, teachers need to encourage open mindedness within the peer group and among parents, teachers and other adults together with the recognition that girls can pursue scientific and technological activities just as well as boys.

2 *The avoidance of gender-specific choices in the choice of pupil projects*. If pupils are given the choice, say of diversity, a device to stimulate colour perception in small children and a device to warn car drivers of over-limit speeds, then it is likely that a clear gender division may emerge. But of course there is no reason why boys and girls should not tackle either project; both are entirely suitable. It is the teacher's task to ensure that there is genuine spread across traditional gender expectations and to ensure that both boys and girls may approach either project with confidence unimpeded by negative expectations – either their own, their teacher's or their classmates'. Particular care is needed when the teaching is being done by a teacher with a gender image – the 'girls' ex Home Economics teacher' or the 'boys' ex CDT teacher'. It may be best if projects were not specifically linked to any teacher and that care is taken to ensure that all teachers in the Technology team are well prepared to teach boys and girls equally. This is not of course to say that the responses of the boys and girls should be indistinguishable. Boys and girls can each bring gender perspectives and solutions to projects that are positive and enhancing and different. The important element is that their choice of project and the evaluation of it should not have pre-determined gender constraints.

3 *Teacher–pupil relationships and the hidden curriculum of expectations and assumptions*. These are a crucial element in perpetuating gender differentiation. Judith Whyte in her study *Girls into Science and Technology* (1986) (a GIST project) has a long list of examples of coded and less-coded examples in which teachers convey expectations of gender differences. She reports:

The gender division of pupils created by separate seating arrangement is frequently exploited by teachers as a mild form of control: [*In Home Economics*] the teacher said to the boys [*in a separate part of the room*], 'Look at the girls over there, they've finished!'

Yet as a means of persuading children to do something, a teacher's instructions weighted with a 'gender signal' can have the opposite effect. By accentuating the behavioural difference between the sexes she is inadvertently teaching the boys not to tidy up, i.e. not to be 'like the girls'.

A craft teacher jocularly called curly-headed boys 'Matilda', ridiculing them in effect for looking like girls. In another school, a misbehaving girl was told off for her 'boyish' behaviour. Here the unintended message is that such behaviour will in fact be tolerated from boys, but not from girls.

Sex stereotypes extend to class work: a craft teacher tried to make the task of colouring isometric cubes more appealing to girls by suggesting they use 'marriage and love' colours, thus dragging in gender quite gratuitously. Boys and girls are being warned of the teacher's expectation that one sex, but not the other, will be interested in a particular task. Dividing the sexes as a means of control or motivation is ultimately ineffective: if something is said to be 'for the boys' it is automatically assumed to be of no interest to the girls, and vice versa. So every topic labelled as 'gender appropriate' is likely to be immediately rejected by half the class, presumably not the teacher's first intention. Gender divisions constantly rebound on the teacher, by reinforcing boys' naughtiness and girls' unwillingness to participate.

Teachers' attempts to be friendly and create rapport often involve calling attention to gender, even when it is clearly irrelevant to the task in hand:

Teacher helping girls in the workshop says jokingly, 'I don't often get the chance to put my arms round a pretty girl.' [*The girl blushes.*]

[*To girl giving good answer*] 'You're beautiful, you!'

While rebuking the girls, the metalwork teacher adopts a softer more pleading tone of voice: 'go and do what you're told', with a rising inflection at the end of the sentence. This contrasts with the monotone excla- mation 'Stop it, lad!' directed towards boys, in a more abrupt and aggressive tone.

'Isn't she a neat writer!'

When girls ask why she has to wear safety goggles, teacher replies, 'You want to stay beautiful, don't you?'

These well-meant asides make children more conscious of gender, and incidentally, of the associated differential expectations that the teacher has for boys and girls: boys must be dealt with firmly, even aggressively: girls can be flirted with, to 'encourage' them along. Boys will be messy and careless, girls are 'neat writers'. The expectations quickly become self-fulfilling prophecies.

None of these incidents alone is likely to be highly significant in changing the perceptions of the boys or the girls. But cumulatively their effect is almost certainly strong and reinforcing.

4 *Special projects*. A number of further projects arc cndeavouring to promote ways in which Technology may be 'more girl friendly'; notably among these is the WISE Project (Women into Science and Engineering) which, although more closely linked to science and applied science has important links with technology. Launched by the Engineering Council and the Equal Oppor- tunities Commission in 1984, WISE speaks directly to girls, arguing, 'Women can no longer afford to ignore the challenging opportunities which careers in science and engineering open up for them'. Central to the WISE strategy is the WISE bus which, staffed with trained teachers, visits schools throughout the country to provide short courses, designed to give girls confidence in their ability to study to contemplate careers in Science and Engineering. Like every other attempt to change fundamental cultural norms the individual effects of the WISE bus may be small but, as a catalyst, it is almost certainly potent and powerful. WISE runs a wide range of awards, courses, visits and other initiatives to encourage girls and women to consider careers in Science and Engineering, they are listed in

Women into Science and Engineering, Awards, Courses and Visits (1991).

5 *Working in the community*. There is little gain in enhancing girls' desire to qualify and enter the construction industry, or boys' desire to qualify and enter the fashion industry if the desire is seen as imappropriately unfeminine or unmas- culine by parents, the community or ultimately by fellow workers and employers. Raised eyebrows, suggestive comments, 'is he a bit queer?', endless wolf whistles, exaggerated courtesy – all these are the mechanisms of social pressure, far stronger than the carefully re- searched and reasoned approach of teachers applying cold logic to these powerful emotive forces. And, as most children know, there are many work situations that are virtually closed to members of the 'wrong' sex – the building site and the typing pool are but two well-known examples.

Again the school can only act marginally to change these social conditions. But teachers can seek to change the minds of parents and em- ployers and identify more areas where girls and boys have succeeded against the expectations and publicise the examples. Above all they can strongly support their pupils who are breaking the gender mould both during and after school.

The path for schools is not easy – the saga of Emily Holt, now a pupil at Bryanston School which has a spectacular new Technology facility was reported by Maureen O'Connor in the *Times Educational Supplement* (11 January 1991) under the headline, 'Bryanston indepen- dent school has a one million pound architect- designed technology building. But almost all the pupils who use it are boys'. She wrote:

> Bryanston has also recruited Emily Holt to its sixth form. Emily was the 11-year-old who went to the comprehensive Camden Girls' School in north London and found that she could not do CDT. Being unusually determined, she made a loud and ultimately productive fuss with a capi- tal F and persuaded John Butcher, then Minis- ter at the Department of Trade and Industry, to help equip a Portakabin workshop at Camden.
> So Emily got CDT on to the timetable. But

five years later her B grade at GCSE turned out to be the end of the road at Camden. Still intent on becoming an engineer and discovering that she could not do technology at 'A'-Level, she made an even more spectacular fuss.

As a result, she obtained what she had set out to find – the financial backing to attend a school which offered Design and Technology 'A'-Level.

At Bryanston all is not quite as well as she expected. She has excellent facilities and teaching, but is the solitary girl in her year in a department dominated by boys. All the new investment, says the head of department, has made technology a popular GCSE option, but of the 56 who will take it next summer, 55 are male.

Mr Wheare, Headteacher, admits there are problems in persuading girls to fulfil their potential in science and technology. Whereas in an all-girl school, all departments compete on equal terms for female custom at options time, in a mixed school, he says, some departments may be happy enough if they can make up their numbers almost exclusively from one sex. 'We have managed to persuade some girls and some pupils you would not describe as high-flyers to do physics,' he says. 'Technology so far has proved more intractable.'

Faced with examples such as these the magnitude of the task is clearly visible. It is all too easy for schools to feel overwhelmed with the situation and to resign themselves to await changes in society which will facilitate this task for them. But to do so is to abandon the immense potential of the schools' contribution to change and to condemn future generations of schoolchildren to restricted opportunities based upon unnecessary and irrelevant gender differentiation.

Racial issues

On the face of it, the existence of racial differentiation in schools is a very different matter from gender differentiation. But the outcomes in terms of effects on children that are based upon irrelevant and unnecessary expectations is very similar. The linkage has been identified by a number of writers, notably Banks (1987).

National Curriculum Technology is particularly sensitive to the way issues of culture and race can be incorporated into a rigorous curriculum and also be firmly based within the educational objectives of the subject.

The non-statutory guidance for Technology quotes these passages from the Design and Technology Working Group Final Report (DES, 1989):

> Cultural diversity has always been a feature of British life . . . [providing] a richer learning environment for all . . . the teaching of design and technology will require perceptiveness and sensitivity from teachers [to take account of] different beliefs and practices, especially when food, materials and environmental designs are involved . . .

And it goes on to observe:

> there are rich opportunities here to demonstrate that no one culture has the monopoly of achievements in design and technology.

The Report urges that this be emphasised as much in white as in culturally diverse schools affirming that Technology has 'an important part to play in preparing pupils for life in a multicultural society'.

However, few studies have been undertaken on the ethnic dimension of opportunity in Design and Technology. But much has been written on the ways in which ethnic minority children perceive their ability, experience and generally lower level of achievement in schools than their fellow pupils. Gillborn (1990) has usefully reviewed the literature and shows that the differential expectations there that so powerfully affect girls' and boys' achievement are matched by similar expectations that determine the experience of young black people and that their expectations are, if anything more persistent than those surrounding gender. At the time of writing (Spring 1992) there is some encouraging evidence of equal or even higher achievement by black seven year olds in the first full round of testing at Key Stage 1. But the evidence is inconclusive and refers to English, Mathematics and Science, not Technology.

It may be argued that a rapidly growing subject area such as Design and Technology could and should provide not only equal but enhanced opportunities for young people who have, so far, not found it easy to 'make it' in the more traditional areas of the curriculum. In particular it should present opportunities for young black people to compete on more equal terms with white children for the new examination, higher education and career opportunities being opened up through Design and Technology.

Certainly some progress has been made. Black children have full access to Design and Technology classes; there is no sense in which their participation is excluded. Yet sadly, ethnic disadvantage in access to examination classes and examination successes seems, in many if not most schools, to be just as marked in Design and Technology as it does in the others. In a subject area in which long-established 'success paths' do not exist to inhibit 'outsider groups', in which formal written English is relatively less important and in which creativity, flair, imagination and style are at a premium, why are black children not able to compete on at least equal terms? Why, one may even ask, are they not sometimes at an advantage?

In suggesting answers, one begins to assemble a picture in which children may be disadvantaged in all aspects of schooling – not just specific aspects. Perhaps at the simplest level of explanation may be the latent, even benign, but unmistakably racist orientations of a school culture which spreads over into the Design and Technology class. Recent visits to a range of Design and Technology classes by the author yielded ready examples:

[*Teacher to class.*] 'This is a messy job and you'll get your hands dirty . . . but that won't worry you Winston, will it?' [*White majority pupils smile, Winston is clearly discomforted.*]

[*Teacher to a group of black boys.*] 'Now then you lads, come down off the banana trees and get on with your work.'

[*Teacher to a group of black girls using an acid bath.*] 'Now then girls put your aprons on please; you don't want to mess up your grass skirts do you?'

There is no evidence that these teachers were intentionally or consciously racist. In discussions with these teachers after the incidents it was clear that they saw it as part of the everyday banter of the classroom, much the same as the banter of the pub, the sports club, the shop or the factory. Moreover some teachers even claimed that it was 'their duty' to expose their pupils to 'the reality of the world outside'. Such remarks were not confined to Design and Technology teachers. On investigation, they were characteristic of most of the staff of the schools. Yet the effects of their remarks were unmistakably playing a part in generating and perpetuating racially linked assumptions for black and white pupils as the response of both groups showed clearly. There was some evidence that school anti-racist policies, where they existed and were implemented, reduced but did not entirely eliminate this kind of teacher behaviour.

Perhaps a more fundamental problem lies in the selection of children for examination streams. Research has shown that black children are often less likely to be given a place in the examination streams leading to high-status 'negotiable' qualifications. Even able black children are denied access because of beliefs that they will lack motivation, be handicapped by language, lack the appropriate cultural backgrounds, fail to understand the system, not 'know how to work hard', or will have behavioural problems and be disruptive. Many of these beliefs, alas, have a tragically self-fulfilling effect (Eggleston *et al.*, 1986).

Design and Technology teachers, like other teachers do not always challenge these assumptions: but they may fail to do so for a special reason. After years of being seen as offering a convenient haven for the pupils defined by the school as non-academic, non-motivated or otherwise of low status, Design and Technology teachers, encouraged by the newly achieved status for the subject, are now enthusiastic to recruit the academic, high-status pupils to their courses. And, sadly, in many schools few black children are yet perceived as being in that category. As one Design and Technology teacher put it, 'after years of having to accept the low-status kids why should we continue to do so now that we can have the high-status ones for the asking?'

A third, and perhaps most insoluble problem in the short run, is the attitude of black parents.

They, often more keenly than white parents, want their children to 'make it' in the educational system. The motivation is understandable; they know that for their children educational success is even more crucial than for many white children because they have far fewer alternative paths to achieve adult success. Not surprisingly, therefore, they view with suspicion new and alternative courses, as yet unproven in occupational and higher education success, and fight zealously for places on the traditional high-status academic courses.

Such parental suspicions are reinforced by the experience of many young black people on Employment Training schemes where they have frequently been disadvantaged in the access to the more attractive schemes based on employers' premises and thereby to a range of job opportunities. This has happened because scheme managers have failed to challenge the racial prejudices of powerful and influential local employers.

For these and perhaps other reasons, we are witnessing a failure to grasp a new and real opportunity for black children to 'make it' in the education system – a failure symbolised by the worryingly low proportion of black children in Technical and Vocational Education Initiative (TVEI) schemes in many parts of the country. This was confirmed by a study of ethnic minority take up of the Technical and Vocational Education Initiative conducted for the Training Agency (Eggleston and Sadler, 1987) in which the preferences of black pupils for 'main stream courses' was very clearly demonstrated. In part this preference was reinforced by poor communication between school and home. Many parents assumed that the work placement for children during the TVEI course was an indication of the job they were likely to get after completion of the programme. Many of the placements in jobs like canteen assistant and supermarket shelf filler were seen as inferior and unacceptable. Similarly, schools failed to explore the extent of parental ambition and support – with Saturday school attendance, encyclopedias and books in the home and much else.

Design and Technology teachers who have played a major role in delivering TVEI must now take urgent steps to work with parents, pastoral and careers guidance colleagues and community leaders to overcome the problem of delivering National Curriculum Technology to black pupils before yet another manifestation of disadvantage becomes reified. Fortunately some encouraging strategies are being developed. One promising initiative is the National Curriculum Council's 'Initiating Activity Through Other Cultures' which is a component of the BBC Television Programme *Design and Technology 11–16: Ways Forward*. The programme notes (NCC, 1991) state:

> Attainment Target 4, level 5 requires that pupils should be able to:
>
> Understand that the artefacts, systems or environments from other times and cultures have identifiable characteristics and styles and draw upon this knowledge in design and technological activities.
>
> The staff at the Holyrood School, Somerset, addressed other cultures and other times. Being a rural school, staff felt that pupils had little experience of the needs and values of other cultures. To broaden pupils' understanding, visitors from Tanzania and India spoke to all Year 7 pupils and gave displays of Indian dancing.
>
> Following this presentation, pupils studied and evaluated clothing, textile design and food products and authentic musical instruments were made from plastic tubing. The effectiveness of these outcomes was assessed in many ways. Food was eaten and discussed in relationship to taste, visual appeal and nutritional value; clothing was modelled by pupils and evaluated with consideration of colour, style, cultural requirements and manufacturing processes. The sounds created by the musical instruments were adapted by the BBC to create the musical composition accompanying this programme.

This example opens up a prospect of opportunities still only incompletely recognised in many schools. It involves a recognition of the rich cultural tradition in Design and Technology that many ethnic minority children bring from their families and communities. Properly recognised in schools it could enable many black children to experience high achievement in Design and Technology and, as the Holyrood example shows, to obtain recognition from their fellow pupils in the process.

Fortunately this recognition is now occurring in other subject areas. Many technology teachers will be helped by a publication of the Association for Science Education entitled *Race, Equality and Science Teaching* edited by Thorp (1992). (It contains an impressive range of anti-racist teaching strategies.)

Conclusion

We have reviewed some of the ways in which the experience and achievement of pupils in Design and Technology is restricted on grounds of gender and race. In general these restrictions are arbitrary and bear no necessary relationship to the basic ability, capacity and aspiration of the individual. However, the self-fulfilling prophecy in all aspects of education is very powerful and it is all too easy for teachers, employers, parents and pupils themselves to create expectations which make the differences seem real and justifiable. Constant vigilance is needed by teachers. However, the National Curriculum requirement to commence Design and Technology at the age of five gives them the opportunity to challenge more readily the assumptions that are often firmly in place by the time secondary education begins at age eleven. Similarly, the integration of 'boys' and 'girls' subjects into a coherent Technology also gives teachers a new window of opportunity through which to build real equality of opportunity.

References

Back, J. A. (1991) 'Multicultural education: its effects on students' racial and gender role attitudes', in J. P. Shauer (ed.) *Handbook of Research and Social Studies in Teaching and Learning*, pp. 459–69. New York: Macmillan.

Banks, J. A. (1987) *Teaching Strategies for Ethnic Studies* (4th edn). Boston: Allyn and Bacon.

Brown, C. (1991) 'Key factors to encourage the participation of primary school girls in Science, Design and Technology'. *Design and Technology Teaching*, Vol. 23, No. 3.

Catton, J. (1988) 'Girls into CDT – some teacher strategies for mixed groups', in J. Eggleston (ed.) *The Best of Craft Design and Technology*. Stoke-on-Trent: Trentham.

Delamont, S. (1980) *Sex Roles and the School*. London: Methuen.

Department of Education and Science (1989) *Design and Technology for Ages 5–16* (Final Report of the Working Group on Design and Technology). London: HMSO.

Doornekamp, G. (1991) 'Gender differences in the acquisition of technical knowledge, skills and attitudes in Dutch primary education: the need for technology education'. *International Journal of Technology and Design Education*, Vol. 2, No. 1.

Eggleston, J. and Sadler, E. (1987) *The Participation of Ethnic Minority Students in TVEI*. Sheffield: Training Agency.

Eggleston, J., Dunn, D. and Anjali, M. (1986) *Education for Some*. Stoke-on-Trent: Trentham.

GATE (1981) 'Objectives of Design and Technology Courses: as expressed in public examination syllabuses and assessment'. GATE Project Report, 1981.1. Chelsea College, CSME, London.

Gillbourn, D. (1990) *'Race', Ethnicity and Education*. London: Unwin Hyman.

Harding, J. (1988) 'CDT – what's missing', in J. Eggleston (ed.), *The Best of Craft Design and Technology*. Stoke-on-Trent: Trentham.

Kimbell, R. *et al.* (1991) *Assessment of Performance in Design and Technology: Final Report*. London: HMSO.

National Curriculum Council (1991) *Design and Technology 11–16: Ways Forward*. York: NCC.

Smithers, A. and Zientek, P. (1991) *Gender, Primary Schools and the National Curriculum*. London: Engineering Council.

Thorp, S. (ed.) (1992) *Race, Equality and Science Teaching*. London: Association for Science Education.

Whyte, J. (1986) *Girls into Science and Technology*. London: Routledge.

WISE (1991) *Women into Science and Engineering, Awards, Courses and Visits*. London: Engineering Council.

CHAPTER 6

Managing Design and Technology

This chapter considers the new management tasks required to deliver Design and Technology effectively and the resources available to facilitate them.

The Technology department of a secondary school is a major cost centre and Technology may well be its largest sector. In a large Technology department of, say, ten teachers with technicians in support, the school budget may well approach half a million pounds. If the interest on the capital cost of the premises occupied is included then the sum may be very considerably in excess of this amount.

In return for these substantial sums very much is expected of Technology. Specifically this includes:

1 Responsibility for 10 per cent of the curriculum.
2 Effective teaching in a demanding subject area.
3 Delivering a co-ordinated programme springing from Home Economics, CDT, Business Studies, Art and Design and elements from many other areas of knowledge.
4 Taking the main responsibility for teaching enterprise awareness both within Technology and also in cross-curriculum activities and, thereby, a major responsibility for ensuring the future of UK industry.
5 Achieving effective liaison with all other subjects of the curriculum.

Arguably, therefore, managing the Technology department is the most difficult management task in the school short of that of the Head Teacher. Its budget is vast, the expectations surrounding it are equally so. And because the content, staffing and resources are so subject specific there is little management expertise elsewhere in the school that can contribute to the specificities of the tasks. These tasks are certainly commensurate with those of senior staff in industry and, as schools become more like businesses, through the local management of schools, opting out and, in increasing numbers of cases, competitive recruitment of pupils, the proximity between running a Technology department and a substantial business becomes ever closer. An HMI paper at the 1989 Loughborough Summer School published in 1990 gave early warning of these changes.

For primary schools the scale is smaller, but the issues are similar and the management task in primary schools is augmented by the problems of introducing Technology to a curriculum unfamiliar with it and in an unfamiliar age range.

It would be easy to present the organisation of Technology as a pure management issue. Certainly all the classic management strategies are relevant and necessary. They include:

1 *Personnel management* – developing an effective and harmonious relationship with colleagues within the department and within the institution generally.
2 *Product management* – ensuring a capability to deliver the product consistently and effectively.
3 *Resource management* – ensuring that the correct support materials, human and otherwise are available at the right time, at the right place and the right quality and within realistic budgets.
4 *Quality control* – monitoring to ensure a consistently high-quality product that does more than barely satisfy legal requirements. It also involves the modifying of standards to respond to technological enhancements and other significant changes.
5 *Client satisfaction* – are pupils, parents and employers sufficiently informed, consulted and satisfied with what is being delivered?
6 *Public relations strategies* – is the image of the institution and its products positive within the workforce, community and, specifically, the client base?
7 *Recruitment and development of staff* – are the most appropriate staff being recruited and retained, are the careers of existing members of staff being enhanced by the appropriate initiatives for promotion and appraisal strategies and in-service training?

Readers will have no difficulty in recognising the relevance of all these activities to managing a Technology department. They are common to virtually every business enterprise and characterise not only the private sector but also, increasingly, the whole of the public sector. This linkage has already been recognised, for example, by the joint scheme of Gloucestershire Education Authority and Unilever plc whereby Technology teachers are being trained in management by Unilever senior executives.

The purpose of the above list is to demonstrate the real management capabilities needed in a Technology department, not only by the Head of the department but all members of it. In the past, management has not always been seen as a priority by teachers of Technology; now it cannot be

ignored and must be done well. Intuitive, flying by the 'seat of the pants' strategies cannot deliver Design and Technology in the 1990s and beyond.

However, the task of this book is not to present general management strategies – many other books already do this effectively. There are, however, some specific tasks of management in Technology, these we will now consider:

1 *To sustain and develop high professional standards using the National Curriculum as a starting point.* This was the theme of Harrison (1990) in which he set out a prospectus for Design and Technology in the National Curriculum. His proposals embody three points:

- The setting of high standards
- The creation of educational rigour
- The development of trust and confidence between teachers and pupils

A project sponsored by the Nuffield Foundation is currently undertaking development along these lines. The achievement of high professional standards in the teaching of Design and Technology is also the object of the Design and Technology Association, which, working in liaison with a range of other organisations, is building a major new professional association to ensure quality delivery of Design and Technology.

2 *To achieve effective curriculum delivery in a subject that has, up till now, been usually delivered by a range of subject departments in secondary schools and without a recognised identity at all in most primary schools.* In virtually every school, such new integration and new identification will usually have to be undertaken by teachers who have at best only familiarity with part of the subject spectrum – often as specialist teachers qualified in one part only or generalist teachers with only limited qualification, if at all, in any aspect.

Many writers have analysed the new organisational patterns schools have adopted to deliver Design and Technology. Perhaps the most useful is Breckon's version (1990). He identifies:

Fig. 6.1 Organisational patterns for the adoption of Design and Technology (Breckon, 1990)

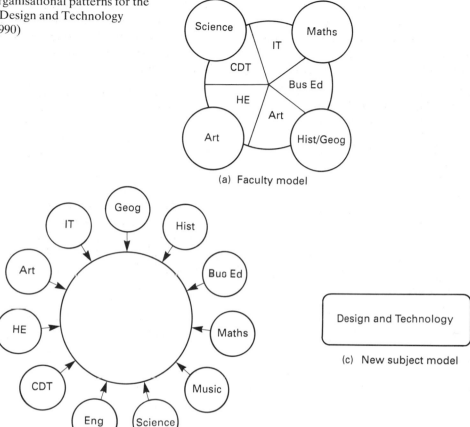

(a) Faculty model

(b) Design and Technology cross-curricular model

(c) New subject model

1 *The Faculty Model* where the component subject departments work together. (Fig. 6.1a)
2 *The Satellite or Cross-Curricular Model* where all subjects contribute to Design and Technology and, out of the combined efforts, Design and Technology capability appears. (Fig. 6.1b)
3 *The New Discipline Model* where all existing subjects cease and a new one is born. (Fig. 6.1c)

The arrangements for teachers to work together tend to take three main forms:

- *Federalist* – where teachers work side by side offering a general range of technology teaching to their classes throughout the year – but with help and support from colleagues.
- *Collaborative* – where teachers may offer their specialist expertise for a period of weeks to a class which is then taken forward by another member of the team in a similar way.
- *Integrated* – where teachers work as a team so that children in any class can call on the expertise of a range of teachers in developing their project and achieving their attainment targets.

Kelsey and Cushing of the National Design and Technology Foundation advocate the role of technological co-ordinators to facilitate these arrangements – but urge caution (*Times Educational Supplement*, 1 February 1991, p. 41):

> The need for co-ordination has been generally recognised. However, headteachers have not always thought through the implications of the appointment, not only in terms of school structures, but, more importantly, in terms of the precise nature of the co-ordinator's role. Few co-ordinators have received training and many are trying to operate in the difficult circumstances of an unclear management brief and only a partial understanding of the technology curriculum.
>
> In one school visited, the problem was highlighted by the premature appointment of a technology co-ordinator some 12 months ago for a team of 19 staff. More recently, realising the important middle management role that a co-ordinator has, the school advertised and appointed a new co-ordinator who will play a key role as part of the senior management team. However, it is not imperative that the co-ordinator is a senior member of staff, providing the senior management team give their full support.
>
> In another school, where planning has a high priority, the co-ordinator works closely with the technology team and senior management in producing a development plan for each key stage. The plan outlines the conceptual and practical skills, in-service and resource requirements. Such plans are essential in order to ensure progression and a clear direction for the team to follow. However, many schools are still a long way from this.

These themes have been addressed by Toft in the Management of Change Project which was based at Salford University; the outcomes are published as a book (Toft, 1989). His findings emphasise the necessity for co-ordination and integration to be seen as key management objectives and demonstrates the careful analysis and utilisation of staff capability that is required to achieve it.

3 *In-service training*. There can scarcely be a school where this is not a management task for the Technology staff. Indeed for some schools much of the work would better be described as initial training for in-service personnel. There are a number of identifiable levels of INSET. They include:

1 Updating of existing Design and Technology teachers in their existing specialism and helping them to locate their work fully in the context of National Curriculum Technology.
2 Extending the range of existing Design and Technology teachers to cover other specialist areas so that they may teach pupils to be aware of them as required to ensure full delivery of the curriculum.
3 Alerting other teachers – particularly heads and colleagues responsible for other subject delivery – to the nature and delivery of Technology.
4 In primary schools, to help all teachers to be aware of the requirements and opportunities of Technology and to find ways in which to contribute to its delivery.

The mechanism by which INSET can be presented varies – through governmental, local authority and school-initiated arrangements. But it is imperative that all opportunities be fully and continuously grasped as the subject develops.

4 *Liaising with industry*. No other school subject has the same urgent need to engage with industry as Design and Technology. We have already mentioned the links between industry and schools in management training for Technology, but here we refer to the more general issue of subject content and its application particularly in such matters as enterprise awareness and wealth creation. A number of specific schemes exist. Caroline O'Grady notes (*Times Educational Supplement*, 1 February 1991):

> General teacher placements are not new. The Teacher Placement Service, which is managed by Understanding British Industry, is a key element to the Government's Enterprise and Education Initiative, aiming to give one in 10 teachers a year a taste of life in the office, or on the factory or shopfloor. There are now teacher

placement organisers in every education authority in England, Wales and Scotland.

A Design Exchange has been set up by the Design Council, the Teacher Placement Service and SATROs (Science and Technology Regional Organisations) to provide opportunities for teachers to work in the design industry on short placements and for designers to work with schools. The scheme is aimed at primary and secondary schools.

Much support for placements and a wide range of information is now available through the Schools Curriculum Industry Project (SCIP). SCIP brings industrialists together with teachers to develop in-school and out-of-school curriculum activities. It is among the pioneers of work experience, work shadowing, mini-enterprises, simulations and other active learning approaches, with an emphasis on enabling pupils to learn from reflecting about their experiences.

SCIP and the Mini Enterprise Schools Project (MESP) were merged on 1 January 1990 to form one organisation committed to the development of a 'work related dimension' for the whole curriculum 5–18. An enlarged SCIP/MESP team includes expertise across the broad range of schools–industry work including enterprise and economic and industrial understanding which has been identified as one of the cross-curricular themes in the National Curriculum. In particular it works in collaborating with the National Curriculum Council Education for Economic Awareness Project in the development of curriculum and professional development materials for 'economic and industrial understanding' and with the Technological and Vocational Education Initiative (TVEI) in the development of models for 'Enterprise in the National Curriculum'.

A good example of the work of SCIP in association with other organisations is the 'Enterprise Curriculum' series (1991) and part 1, Book 7, *Case Studies in Technology* contains a range of examples of school projects. Case Study 3, 'The Brintons' Carpet project' is typical and is reprinted with permission (pp. 74–5).

At the primary school level, much has also been achieved by the work undertaken by Edge Hill College of Education which has demonstrated the feasibility and worthwhileness of industrial experience for very young children. *The Enterprise, Economic and Industrial Understanding Kit for Primary Teachers* from the Primary Schools and Industry Centre at the Polytechnic of North London is another valuable resource. And of course much assistance is available from the National Curriculum Council in its task of pervading the curriculum with the enterprise culture.

The SATROs (Science and Technology Regional Organisations) have a history of enriching the science and technology curricula in schools through the promotion of links with industry and business. In addition, SATROs work with teachers, individually and in teams, in co-operation with people in business and industry, to produce new curriculum materials to support the technology curriculum. These resources range from case studies of successful school–industry link activities, to material, in printed, audio and visual format, which provides starting positions for technology projects at all key stages. These materials are published nationally by the umbrella body for SATRO network, the Standing Conference on Schools' Science and Technology (SCSST).

'Much of the case study material has been, and will be generated by the SATRO Learning from Industry project. Funded by the DTI, Learning from Industry enables teachers to be placed with local companies for the specific purpose of generating curriculum resources based on that company's products, processes or services. As a result of the experience gained in placing teachers with companies for this and other projects, SATROs will work with the Teacher Placement Service (TPS) to provide briefing and debriefing sessions for teachers of technology undertaking placements through the TPS.

In addition, many SATROs offer INSET based on their local expertise in technology and the use of industry links. Some of the INSET offered will be based on the nationally published curriculum materials, others on such SATRO

Project: Brintons' Carpet Project
School: Holyhead, Birmingham
Contact: Mr D. Taylor, Head of Art and Design
Technology Input: Art/Design, Information Technology, Technology and Textiles
Outcome: The creation of carpet designs for a local manufacturer

The Project

Holyhead has been developing Schools–Industry Links since 1983 to give its students greater awareness of the world of work. Work experience and local collaborations gave rise to the Project. Brintons' Carpets fitted ideally into their cross-curricular plans.

The Project extended over two terms and accounted for 15 lessons × one hour per week for 120 pupils. The task was of GCSE standard for Year 9 pupils, some with hearing problems.

Research

On-going curricular development plans.

Aims and Objectives

   (i)   to develop a carpet pattern
   (ii) to prepare a report on the carpet industry
   (iii) to promote teamwork and,
   (iv) to take risks.

Team Build

A number of preparation activities from the assessment file were used for team-building exercises.

Action Plan

To produce an end carpet design within a research framework. The activity was run as a competition. In particular, the process focused on developing ideas on paper. The task was completed by presenting the work to Brintons using media and IT skills.

Implement Action

The students were given multi-cultural briefs and set themselves in teams to work. They were introduced to their responsibilities within a framework of active learning and simulated industrial reality. Brintons was visited and experiments carried out with dyes. The students possessed considerable freedom within their goals.

Review

The competition ended with assessment of the group designs although continuous assessment had occurred as the pupils kept log-books of their enterprise competencies. A report was prepared by the teams that summarised their feelings for what they had done.

Staff Comment

They took the risk of 'letting go' and pupils responded to the challenge. The pupils called teachers by their first names as they would have done in a real work-place. This was not abused and gave an atmosphere of authenticity and real working relationships.

Student Comment

They commented: 'We had to develop the skill of being more aware of barriers that exist.'
They were able: 'To make decisions and be in control' and emphasised 'the importance of students taking direct responsibilities.'
The Project gave them a clear idea that: 'Hard work, determination and devotion are definitely the keys to success.'

National Curriculum Attainment Targets met

The students' comments reflected one overriding quality – surprise in their own abilities. Further, it highlighted the challenging nature of the new, exciting learning demands.
The Project permitted a wide range of AT delivery.

National Curriculum Attainment Targets met

ATTAINMENT TARGET 1
Identifying needs and opportunities. Level 4e, f; Level 5a, b; Level 6a, b, c; Level 8a, b, c; Level 10a

ATTAINMENT TARGET 3
Planning and making. Level 4c, d; Level 5d; Level 6a, b, d, e, f; Level 7a, b; Level 9a, b; Level 10a

ATTAINMENT TARGET 4
Evaluation. Level 6a

ATTAINMENT TARGET 5
IT Capability. Level 6a, c

Enterprise Attainments in Technology met

Cross-referencing with the Curriculum Map on page 8 shows considerable coverage of Technology ATs by this enterprising activity.

Potential

This Project led on to RSA accreditation and a visit by the Design Council to the school. A travelling exhibition later used the winning designs. It makes clear the obvious advantages of 'adopting' a local firm and the possibilities schools may develop for themselves in their own localities.

Preparation and Useful Tips

- Make strong ties with local firms and carry out pre-project visits to run through the programme and check the logistics yourself.
- Bring the contacts at the firm into the school and the classroom.
- Check safety requirements and your insurance cover for visits.
- Tell the teams they will be assessed on their teamwork as well as the outcomes.

---

activities as the CREST Award Scheme, a joint project of SCSST and the British Association. CREST with its basis in scientific and technology project work, often with industrial involvement, provides a good vehicle for the delivery of the technology curriculum. For over 12 years SCIP has been promoting links between industry and education. The value of such links has a particular relevance for technology in ensuring that the context in which Design and Technology activity takes place includes situations such as the home, school, recreation, community, business and industry.

Yet though most Design and Technology teachers are aware of the importance of offering commercial, economic and enterprise under-standing there is often an unwillingness to consider the real world of business with its emphases on matters such as cost efficiency, added value and profitability. Schools frequently focus projects on caring, welfare and environmental projects rather than on mainstream 'wealth creating' commercial activity. Even projects on retailing and marketing are not infrequently based on charity shops!

5 *The organisation of the Technology area.* Design and Technology is not taught in classrooms but usually in specially designed and equipped laboratories, suites, workshops and studios. Even in schools for very young children, special equipment and facilities are increasingly available and in use. Management of these requires

specific considerations. They include, for instance, the crucial matter of safety in the use of dangerous machinery and hazardous materials, dealing with dust and other forms of waste, eye protection, and much else. The development of effective guidelines, storage systems, etc. is vital and particularly so in primary schools where the problems are new and the activity may be taking place in general rather than specific purpose areas.

Safety is but one of the components of organisation. The identification, application and maintenance of resources is a major task and the need to keep up to date with an equipment market which is developing at least as rapidly as the subject area itself is essential. Technology teachers are unlikely to be short of information – most teachers' journals feature extensive resource information and in addition there is the flow of catalogues, bulletins, literature which reaches torrent dimensions in many schools. The objective evaluation of this material – much of it very persuasive – is a key requirement and the risk of making a faulty decision on restricted budgets dominates the minds of many teachers. Industry has its purchasing officers but no school Technology department has yet reached this level of sophistication. The responsibility is heavy and daunting for many teachers.

Of particular importance are the books and videos. Some impressive new series of pupils' books are available or in preparation; those from the Oxford University Press, the Cambridge University Press and Longmans, amongst others, are very promising; it is important that these are distinguished from updated versions of established texts in the contributory subjects. Excellent video material is now available notably the series from Thames Television, Seeing and Doing (Key Stage 1), Designing and Making (Key Stages 2 and 3), Design and Technology (Key Stages 3 and 4) and Business Studies (Key Stage 4). There is also the BBC's award-winning Teckno Series (Key Stages 3 and 4). It is also desirable that at least one member of the Design and Technology department attends the Design and Technology Exhibition at the National Exhibition Centre in October each year if only to see the sheer magnitude of the resources available and the attention that a wide spectrum of industrialists and the public attach to it. Above all, the management of the area must focus on quality – of environment, resources and teaching so that pupils receive the optimum technological experience.

6 A further issue is timetabling. Unwin (1990) points out some of the issues clearly:

> Without doubt schools will need to make broad estimates of time to be given for the different areas of the curriculum. This is not a new concept, though the issue is not so clear in relation to design and technology. Picture a class of mixed ability children and think of the varying times which groups require to complete a task. We all know that the same task can require a wide range of allotted time across even one class. The very nature of design and technological activity will provide situations in which groups of children are tackling a diversity of tasks, perhaps related to the one overall problem, with the obvious consequence that some children will have completed their task whilst others are still deep in activity. Design and technology activity is not like any other areas of the school day whereby work continues largely on an ongoing basis.
>
> A simple solution would be to give groups a fixed period of time in which to complete their task at the end of which they stop working no matter whether they have finished or not. The effect on a child's morale using this approach is only too dreadful to imagine, especially when the group is committed and motivated to solving the task. A different approach would involve the greater use of the teacher's professional discretion with attempts to maintain a balance over a whole school year. In practice, I believe this is asking too much of a class teacher. Another alternative is to clarify with the children the idea that time is a resource akin to the other resources they might be using and perhaps to qualify their suggested solutions before continuing with their work. This might mean a teacher reducing the scale of a group's suggested plan. There does not appear to be a simple answer to this issue but it should be at least the

basis for discussion within each school so that there is a likelihood of equality of opportunity over a child's primary school career. Unfinished projects however are surely unacceptable to us all.

Unwin goes on to discuss the problems of group size and organisation and with tongue in cheek presents the scenario for a whole school policy:

Picture a school, parts of which might be depicted as follows:

- **Red Class:** Children all working together making a model trolley to carry bricks from one end of their (imaginary) building site to the other. Thirty children all using scissors, glue and endless resources at the same time. One group has been trying to get the teacher's attention for twenty minutes but without success.
- **Green Class:** A small group of three children are in the activity corner cutting card to match those parts of their design proposal which is pinned for them on to a sloping board close to where they are working. Large, amusing but clear pictures surround this corner reminding children of how to 'behave' when engaged in activity in this area of the classroom. Another group are busy finishing their design whilst other groups are engaged in other work.
- **Blue Class:** The children are very excited by the prospect of making the trolleys they have designed but when they look in the found materials box (the contents of which they had to tip on to the floor to look for what they wanted), they found that the previous groups had used all the cardboard wheels. Their teacher had also forgotten to bring the art straws from the stockroom at the other end of the school and they were not allowed to get their own materials unless it was in the found materials box.

Obviously, this sort of scene is not likely to be found in any school but it shows that a whole-staff approach would have allowed the weaker practitioners to benefit from the good practice of other teachers.

Quite apart from the subject specific issues in classroom management there are more general issues of justice and fairness which it is crucial

for Design and Technology to examine. We have already glimpsed some of these issues of equality of opportunity in the previous chapter on gender and ethnic issues.

7 Finally, we must mention the whole range of support organisations that can facilitate the management of Design and Technology. We have already noted the professional organisations now largely grouped under the Design and Technology Association which provide an increasing range of support for training, resource selection, publication, conferences and advice generally. The magazines of the Design and Technology Association, *Design and Technology Teaching* and *Primary Data*, demonstrate a good example of such information delivery whilst the *International Journal of Technology and Design Education* offers a regular international perspective of developments in the subject. Other relevant professional associations such as DESTECH and the Independent Schools CDT Association work closely with DATA.

The role of the Design Council is of considerable importance. The Council provides a wide range of support materials for teachers and students with permanent and temporary exhibitions and a publication *The Big Paper* specifically designed for schools. Yet another important support organisation is the National Council for Educational Technology providing a wide range of courses and support material.

There is also the Design Dimension Project promoting services for teachers, which, in liaison with the Design Museum, offers a range of new courses for primary and secondary inservice. Each course offers high-quality support materials which teachers can use in schools. This is part of a National Curriculum development and training initiative for primary and secondary schools being developed by a Design Dimension/Design Museum Partnership and a consortium of Local Education Authorities.

There are also the competitions which are well established such as the Young Designer Competition sponsored by Toshiba and with the support of the Confederation of British

Industries, and the highly successful Young Electronic Designer Awards, sponsored by Texas Instruments and Mercury Communications – these and many more provide powerful incentives for primary and secondary schools and also follow through into Higher and Further Education.

The British School Technology Trust, through BSTE (Limited) offers a wide range of support, particularly in the design of teaching areas – both static and mobile. The Trust's brochure notes:

> Recent curriculum work has included IT audits, the development of secondary phase cross-curricular assignments and evaluation of the 'Technology for All Across the Curriculum' (TAAC) project. Consideration of curriculum issues lies behind all BSTE educational activity.
>
> Staff development programmes include residential INSET at Carlton varying in length from two-day short courses to intensive schedules of six weeks duration. A wide range of Design and Technology programmes have been presented to groups of teachers from all phases of education. These have varied from primary 'Getting Started' to advanced level 'Control Technology' courses. Day courses are also presented to BSTE lecturing staff at suitable venues specified by our clients.
>
> BSTE offers advice and guidance to schools and colleges willing to introduce Design and Technology and can if required provide a full implementation package including the supply and installation of furniture, equipment and services. BSTE keeps extensive records of suppliers of relevant resources and provides manufacturing capability to satisfy requirements not readily or adequately provided for. If requested by the client establishment we are happy to liaise closely with architects so that we can jointly ensure that both modifications to existing buildings and the design of new environments meet the educational requirements appropriate to that school or college.
>
> The BSTE team are also responsible for the design and conversion of vehicles into Mobile Environments, a large number of which have been dedicated to Design and Technology education.

Yet another source of information and support is through NERIS (The National Education Resource Information Service) which in conjunction with Nottingham Technology Education Development Group (NTEDG) can supply information in support of National Curriculum Technology. NERIS is an independent educational trust which provides a computerised information service of educational resources and developments for teachers and pupils in schools for all age groups. Schools, colleges and other educational establishments throughout the UK gain access to NERIS either through an on-line service or on CD-ROM.

Perhaps most important are the publications of the National Curriculum Council – interpreting, augmenting and exemplifying the legal requirements of the Education Act 1988 and demonstrating the immense potential for teacher initiative that exists within and beyond a National Curriculum.

Conclusion

Management of Design and Technology is justifiably a vast subject which, like the subject itself has grown rapidly. It ranges from mega-issues of the kind encountered in large commercial enterprises to micro-points of detail which are still crucial in the effective quality delivery of the subject – so that achievement may be fully and fairly available to all children. Fortunately there is a wide range of support for teachers in managing Design and Technology. The range of statutory and non-statutory agencies is extensive – teachers are not alone.

This chapter has argued that Design and Technology teachers must see management as a crucial part of their role – to be pursued actively rather than passively. It is part of the design process of delivering technology – the delivery of technology requires a technology of delivery.

References

Breckon, A. M. (1990) 'Design and Technology in the National Curriculum'. *Design and Technology Teaching*, Vol. 22, No. 1, pp. 4–11.

Harrison, G. (1990) *Towards the Implementation of Design and Technology in the National Curriculum*. Bournemouth: ICHF.

Kelsey, B. and Cushing, S. (1991) 'Nuts and bolts and know how'. *Times Educational Supplement*, 1 February, p. 41.

Loughborough Summer School (1990) 'Managing CDT Departments in Secondary Schools in the Context of LMS'. *Design and Technology Teaching*, Vol. 22, No. 1, pp. 19–22.

O'Grady, C. (1991) 'Tasting the real world'. *Times Educational Supplement*, 1 February, p. 42.

Pennell, A. and Alexander, P. (1990) *The Management of Change in the Primary School*. Lewes: Falmer.

Schools Curriculum Industry Project (1991) *The Enterprising Classroom, Part 1, Book 7, Case Studies in Technology*. Coventry: SCIP.

Toft, P. (1989) *Managing Change in the CDT Department*. Stoke-on-Trent: Trentham.

Unwin, M. (1990) 'Classroom organisation'. *Design and Technology Times*, pp. 14–15. Salford: The University.

Useful Addresses

Annual Design and Technology Exhibition (National Exhibition Centre), I.C.H.F. Limited, Dominic House, Seaton Road, Highcliffe, Dorset BH23 5HW

BBC Education, TV Centre, Shepherds Bush, London W5

British School Technology (BSTE), Carlton, Bedfordshire MK43 7LF

Central Independent Television, Central House, Broad Street, Birmingham B1 2JP

Centre for Alternative Technology, Machynlleth, Powys SY20 8DN

Council for Environmental Education, School of Education, University of Reading, Bulmershe, Reading RG3 5AG

Design Council, 28 Haymarket, London SW1Y 4SU

Design Dimension Educational Trust, Dean Clough, Halifax, West Yorks HX3 5AY

The Design Museum, Butlers Wharf, London SE1 2YD

Design and Technology Association (DATA), 16 Wellesbourne House, Wellesbourne, Warwickshire CV35 9JB

Design and Technology Education Research (DATER), John Smith, Department of Design Technology, Loughborough University of Technology, Loughborough LE11 3TD

DESTECH, 24 Woodland Drive, Sandal, Wakefield WF2 6DD

The Directory of Local Contacts for Business and Education Partnerships, Janet Jones Associates Limited, Queen's Building, Westfield College, Kidderpore Avenue, London NW3 7ST

National Association of Teachers of Home Economics, Hamilton House, Mabledon Place, London WC1H 9BJ

National Council for Educational Technology, Sir William Lyons Road, Coventry, CV4 7EZ

National Design and Technology Education Foundation, 45 Bournemouth Road, Chandler's Ford, Eastleigh, Hampshire SO5 3DJ

NERIS/NTEDG, Maryland College, Leighton Street, Woburn, MK17 9JD

Open University Central Enquiry Service, PO Box 71, The Open University, Walton Hall, Milton Keynes MK7 6AA

Primary Schools and Industry Centre, School of Teaching Studies, Polytechnic of North London, Prince of Wales Road, London NW5 3LB

SCIP/MESP, Central Office, University of Warwick, Department of Education, Coventry CV4 7AL

Thames Television Schools Office, 149 Tottenham Court Road, London W1P 9LL

Trentham Books Ltd (Publishers of *Design and Technology Teaching, Primary DATA* and *International Journal of Technology and Design Education*), Westview House, 734 London Road, Oakhill, Stoke-on-Trent ST4 5NP

Trent International Centre for School Technology, Nottingham Polytechnic, Burton Street, Nottingham NG1 4BN

Understanding British Industry, Tel. (0865) 722585

Young Electronic Designer Awards Trust, 24 London Road, Horsham RH12 1AY

CHAPTER 7

Design and Technology in practice

This chapter is devoted to examples of Design and Technology practice in Key Stages 1, 2 and 3.

We have discussed the nature of Design and Technology, its provenance, its place in the National Curriculum, its assessment, its availability and its management. This chapter is devoted to presenting examples of the practice of Design and Technology in the schools – reported by the teachers themselves. We include four reports, one each on work in Key Stages 1, 2, and, because of the greater diversity, two on work at Key Stage 3. At the time of writing the amorphous state of Key Stage 4 development makes it impossible to include a representative example of work at this stage. Each example covers all four Design and Technology Attainment Targets and offers work appropriate to the Levels of Attainment applicable to each stage. Each activity described fits happily into the Design and Technology Programmes of Study.

There are important points to make about all the examples. Though all are concerned with Design and Technology they all involve contributions from a wide range of subject areas – and also make contributions to a wide range. Though clearly identified as a component of National Curriculum, Technology work is not compartmentalised but highly integrated with all other subject areas.

A second important aspect is that though all the examples are clearly and directly in accord with the requirements of the National Curriculum, they are

all accounts of projects that are complete in themselves and offer exciting, interesting and satisfying activities for children – and attractive teaching opportunities for the teachers who plan and deliver them. The requirements of the National Curriculum need not and should not distort or diminish the work of the children or teachers; instead they should enhance it.

We begin with Making Musical Instruments at Key Stage 1 by Toni Kane, followed by Developing a Structured Approach to Design and Technology at Key Stage 2 at Marston Green Junior School reported by John Dainty, the Key Stage 3 Technology, Research, Information, Problem Solving Course (TRIPS) at Greenhead reported by G. W. Asquith and D. Smith, and finally, Putting Theory into Practice in Key Stage 3 at Vandyke Upper School, reported by Peter Jones.

There is one common feature that perceptive readers will notice. All projects, particularly those for the older pupils, are predominantly paper projects. At an early stage in developing Design and Technology in the National Curriculum this is hardly surprising – such projects can be developed more readily, adapted more quickly and do not have the prior resource and equipment needs of 'hard' three-dimensional work. Yet many educationalists are concerned lest paper projects become the norm and lead to the diminution or loss of skill in realising technological projects. A strong emphasis of this book is that such realisation is at the heart of Design and Technology and we shall hope, in future editions, to be able to provide a full range of appropriate case studies that build on those presented here.

Meanwhile, a number of books offer guidance on how to develop work in Design and Technology; a typical example is Lever's *National Curriculum Design & Technology Key Stages 1–3* (1990). Such books offer detailed suggestions for teachers on a range of activities that may be suitable and how they might be planned, presented, achieved and evaluated. But that is the task of different, complementary books. Here, in this volume, we let the projects and the teachers present themselves in words, pictures and diagrams. Having read this book so far, readers will be able to understand the situations and contexts in which the teachers are working and will see clearly the imaginative and creative way in which they are beginning to use the opportunities that can be offered by the National Curriculum and the generally enhanced status of Technology as a Foundation Subject – albeit at a still early stage in its existence.

MAKING MUSICAL INSTRUMENTS AT KEY STAGE 1 AT UPLANDS COUNTY PRIMARY SCHOOL
Reported by Toni Kane

During a Summer term I worked with my class of 28 top infants (6–7 years old) on a topic based on Sound and the reduction and transmission of sound. I hoped that this topic would develop children's skills in most areas of the curriculum. I saw the topic falling into the following main categories:

Topic Plan

Discussion of the meaning of sound

What is sound? How do we use it? Which senses are involved? How would it feel to be deprived of sound stimuli? Have all life forms a sense of hearing? Are some better than others? Could we do some experiments to test whether live things can hear or not? How many different ways of making sounds can we think of? Can we use our own bodies? Can we use other humans or live things? Can other humans or live things produce sound, e.g. wind, rain, thunder, fire, water, earthquake, volcanoes? Man-made artefacts such as windmills, engines, sirens, radio receivers, computers, clocks ad infinitum! and so to music or sound for pleasure.

Listening to sound

We should listen to recordings of different types of sound in the classroom. Can we say what is the difference between sound and musical sound? How do we make musical sounds? Listening to specific musical instruments with (if possible) examples of instruments or pictures while listening. Discuss how sounds are made using musical instruments. We should elicit hitting, plucking, blowing, scraping, shaking. Listening to synthetic music and looking at, for example, Casio and other forms of electronic music.

Looking at instruments

We can bring instruments into school and we have a wide variety in school and homes. Particular emphasis should be to elicit from the children how the instruments are made. What materials are needed? Why is measurement important? What does the instrument need to have to make the sound more interesting? (Give example of teacher's whistle compared to, for example, a recorder.) An elementary discussion of pitch using a violin or guitar, recorder or even water in bottles should follow. The children should be able to observe that pitch difference is related to size or volume. The loudness can also be changed, either by putting in more energy or increasing the volume of air around the sound (sound box).

Making instruments

The children should now be allowed to experiment on these aspects using materials in the classroom. Musical instruments may be produced and these could be put on display by children who should organise this themselves, used in music-making sessions, recorded, used in a class assembly. All this work should be recorded in a topic book with observational drawings, plans and descriptive writing. We could make a class orchestra with strings, timpani, woodwind, brass, percussion, etc.

Using the computer

We have a good program on pitch and rhythm quite suitable for this age group. It is called *Jolly Jack Tar* and teaches rudiments of pitch and rhythm. Some children may follow this up by writing their own simple music.

Using an oscilloscope

At this stage it would be useful to see how sound can be visually presented. The oscilloscope can be demonstrated at school.

Art and craft

The instruments should be made to look good – we have to look at them while listening to music. I should also like to try painting to music, i.e. listening to music and painting what we hear.

I hope that the planning, recording and making will all be presented in the child's own topic book and work done displayed in the classroom. I see the topic as encompassing the following areas of the curriculum: English (the discussion and recording of work); Maths (the measurement involved in making; an elementary understanding of pitch and its mathematical implications, i.e. the concept of ratio; ideas about volume capacity and length); Art and Craft (the drawing, making, painting); Design (the designing of instruments); Technology (the use of suitable materials in the manufacture of the instruments); Science (experimentation with sound and ways to produce sound in non-living things – for example, sound on screen – the oscilloscope; mechanical and electronic sound, for example, radios, telephones; PE and Drama (using the body to make sounds, interpreting music through movement); Computer (use of computer to teach pitch and rhythm); Music (enjoying making music, writing music, listening); RE (use of music in religious services, hymns); Environmental Studies (briefly touching on different types of instruments nationally/worldwide, how we can identify a national music, often by instruments used (local materials)); how instruments have developed through history.

Resources and timescale

- Science packs in school on senses
- Model or diagram of ear (Fareham and Gosport Teachers' Centre)
- Recordings of natural and mechanical sound (in school)
- Instruments (in school and elsewhere)
- Visits – possibly West Dean College, instrument makers
- Mechanical Music Collection, Chichester
- Gabriel's Horn House (musical shop), Southsea
- Materials for music making – scrap materials and wood, elastic bands etc.
- Cassette recorder
- Recordings of different national music

The timescale of 15 weeks in the Summer term should be ample to cover this topic very thoroughly. I would envisage 2½ hours per week on this topic. I should imagine giving three weeks to discussion on sound, preparing of topic books, collecting materials and made instruments, listening and experimenting in sound – recording discussions and planning and design of own instrument. (This could be done individually or in groups as I have parent helpers on one afternoon each week.)

The making part of the topic should take 2–4 weeks depending on complexity of instrument.

The music-making part of the topic could take another 3–4 weeks and some time will be given to visitors and visits, plus a possible class assembly which might take a whole week to prepare. This will give room for expansion of any of the aspects of the topic – should the children be needing more time on it, or for a look at the transmission of sound with, for example, the telephone and radio.

The topic review

Having covered most of the ground which was planned in the original layout I now received and evaluated the work. In order to clarify the operations I have decided to itemise the procedures as follows:

1 The preparation and timescale
2 The timetable
3 The visual aids and peripherals
4 The discussion and science work
5 The making of instruments and recording
6 The display
7 The visits
8 Cross-curricular applications
9 The Assembly
10 Samples of children's work
11 Photographs
12 Bibliography

1 The preparation and timescale

Having decided on my topic, I made out a Topic Plan as already submitted. It should be realised that for much of the practical work I was dependent on the help and availability of parents and that if they did not arrive one day then proceedings were accordingly slowed down! Also I had to choose a day when other interests or parts of the curriculum did not interfere, when the parent-helpers could keep a regular date which would not interfere with their own arrangements. I have a long-standing arrangement for parents to come in on Wednesday afternoons so this stood. Another

parent whose youngest child had just started school offered to help in a more flexible way and came in on request through message-by-child! I should also say that the week of the Assembly I had a helper in the form of a student on the Trident scheme and, as luck would have it, she became very involved with the work and was an invaluable helper. It is important to realise that much of the practical work would have been extremely difficult to achieve without the help of non-teaching, and in this case voluntary, assistants.

2 The timetable

Following the timescale for the actual making of instruments. All other activities (for example, discussion, other art work, work about hearing and sound in general) were also going on.

Week 1
Hanging plant pots
Water in bottles (milk)
Xylophone – hanging
Drums
Bongo – drums

Week 2
Claves
Castanets
Xylophone – hanging (continued)
Bongo – drums (continued)

Week 3
Xylophone
Bongo – drums (continued)
Jingle pole

Week 4
Closure on the Wednesday

Week 5
Trip to West Dean College and Mechanical Music Collection, Chichester

Week 6
Tubular xylophone
Jingle pole (continued)

Week 7
Visiting teacher on Work Experience for day

Week 8
Tubular chimes

Week 9
Zither
Scrapers
Maracas
Pan pipes
ASSEMBLY

3   The visual aids and peripherals

TV programme *Seeing and Doing* – schools programme suitable for 6–8 year olds with emphasis on musical instruments and the way sound is produced in them. I have video-taped this series for future reference as 'sound' is to be an infant Science Topic in our long-term plans linked to the National Curriculum. We were entertained by the Peripatetic Music Staff playing violin, viola, cello and piano. For the sound and hearing part of the topic I obtained a model of the ear, showing the middle and inner ear, from the Fareham and Gosport Teachers' Centre. I also made a game using six tins containing various objects. From a list of the six items the children had to make their choice and mark it on a score sheet as follows:

	Tina	Sam	Carly
Drawing pin	4		
Rubber	2		
Marble	1		
Cotton wool	6		
Rice	5		
Dried peas	3		

I used a book of *Noisy Poems* – all with 'sound' words – many of the children have been memorising them. I read the story of *Buttons* – the hearing dog for the deaf – we discussed ways in which an animal could give visual signals to a deaf person. We wrote Sound Poems and sound stories. We covered the story of Joshua and the Battle of Jericho and dramatised it. We listened to records of sounds, for example, traffic, telephones, thunderstorms, etc. One of the staff plays the violin and she came and played and talked about how the instrument is made, tuning, etc. I brought my guitar to school and tuned it while the children watched. Also I played while we sang.

4   The discussion and science work

I started the discussion on Sound by asking the children to close their eyes and concentrate on what they could hear. Some heard a blackbird, a creak, rain, etc. We discussed sound and how it must travel from its source and through the air – we discussed sound waves. How is sound made? We discussed this and then I used a Tasmanian drum (made from clay with a skin over the base and an open top) resembling a vase – the base of course has been removed before the skin is attached. It makes a beautiful resonant sound. However, if we stood the drum upside-down, so that its open end was sealed, the sound became dull. This led into discussion that instruments, not just drums, need a space in which the sound could resonate and also an aperture for the sound to emerge in order to reach the ear.

At a later date we did more experiments. The children closed their eyes and we tested direction by taking turns to clap, click, tap, etc. and guessing in which direction the sound was made. We tried putting cones of paper round our ears, or margarine tubs with bases cut out to see if it had made a difference to how well we heard, but we noted hardly any difference.

We looked at the model of the ear and soon the children understood how sound is passed through the ear by the vibrating of the ear drum and middle ear bones. We talked about *how* you know what sound you are hearing. Sound sources could be animals or inanimate objects like machines, cars, etc. or even weather.

Interest level was high during all discussions.

Having also listened to recordings of various sounds, for example, telephone, thunder, rain, vehicles, etc. we came to listen to some music. What was the difference? Many answers came up. Eventually pitch, rhythm and being played came and we thought about *how* they were played. We came up with hitting, plucking, blowing, scraping and shaking.

Prior to the practical work we talked about

various ways in which we could make, for example, a hitting instrument, and how we could make beaters – what materials would be best and the strongest way to make things. We also talked about using tools (for example, saws and drills) and how to use them safely. (I was rather wary, not having used saws before with an infant class!) I am happy to report that we had no accidents at all as the children felt very privileged to be using tools and took it all very seriously. I had intended to take some photographs of the children working but I always get so involved personally that I either forgot or did not have time!

5   The making of instruments and recording

I decided to keep to a structure where the method of playing as in hitting or tapping was the predominant factor. So we chose from the many resources we had what each group would like to make. The basic idea was on the work card or in the book but the children came up with modifications or ideas of their own with surprising ease. Because of limitations in availability of materials I had to pre-choose what we would make on a certain day. On the first day of making we had a choice of flowerpot chimes, drums and beaters, bongos, hanging xylophone or bottle chimes.

All of these were completed on the day except for the bongos which were made from a papier mâché mould and took several weeks to complete. Interestingly, because the bongos took so long to make, the children found it difficult to record the procedures they had gone through. Also, we struggled to secure a plastic skin on the paper because it tended to bend and buckle with the elastic. We even tried bracing the base with dowel. Eventually a child turned them over and found they sounded better if we tapped the 'bases' so we did not bother to put a skin on them.

The groups consisted of five or six children.

Week 1
- Flowerpot chimes – made from clay pots of various sizes – painted in bright designs, left to dry (powder paint) and hung with flex from a broom handle suspended horizontally. Pots had to be graded for size. Pitch varied with size but not as you would expect. Quality of clay must have had an effect – some pots were dead sounding but I could find no cracks.
- Drums – base made from various cylindrical containers; skins were plastic sheeting from a pram hood. Beaters made with dowel and corks, conkers, etc.
- Bongo drums – took three weeks to complete. A glue gun would have been useful to do some of the fixing (for example, the wedge-shaped piece of wood to join the two drums together was difficult for small children). Papier mâché bases around a mould were very successful.
- Hanging xylophone – here measurement was involved. Children had to decide on lengths of pieces of wood. Standard measures were used and after the longest piece was decided the children worked out how much shorter each other piece had to be. Bench hooks and saws were used, sanding the ends and tying with string.
- Bottle chimes – very easy to set up. Interestingly, the resonance was better when the part with liquid was tapped than when the part with the air was tapped.

Helpers on this session were two parents and a floating teacher, so four adults in all.

Week 2
- Claves – pieces of broom handle cut to same length (measurement involved). Some of the handles were already painted. Sanding of ends.
- Hanging xylophone – attaching to string to hang very difficult.
- Castanets – we made finger castanets with bottle tops and loops of ribbon to hold them onto figures.
- Bongo drums – continuing.

Some of the children are now writing and discussing the making that they have done. They often need help to be reminded of how to record their work. I have said it is like a recipe in cooking (they have a cooking session every week if it is their group's turn). We have to write down the ingredients or materials used, then how we made it

– finished it – and finally how to play it and how it sounded. They most often forgot this last part! However, with some persistence the writing did improve and some of it is very good quality and much better than some of their 'news' or 'story' work.

Week 3

- Xylophone – wooden lengths on a painted shoe box with an aperture in the lid. The pieces of wood are resting on pieces of draught-excluder spaced according to the required distance.
- Bongo drums – continued.
- Jingle pole – another broom handle, beer bottle tops are painted then loosely tacked to the pole. A door stop is nailed onto the bottom of the pole. The pole is banged on the floor to make a jingle sound.

Week 6

- Tubular Chime Bar – one large piece of metal tube is set on a painted shoe box with lid and aperture. Also seven graded lengths as above.
- Jingle pole – continued.

Other groups are recording the making they have done.

Week 8

- Tubular chimes – as tubular chime bar but using several copper pipes of graded lengths. They had to be sawn with help from an adult. These were suspended from a wooden frame.

Week 9

- Zither – shoe box painted and reinforced with plywood glued into lid. Aperture made at home by me. Screw eyes and guitar strings used to make a tuned, plucked instrument.
- Maracas – squeezy lemons, wash-up liquid bottles – glued, papered and painted. Filled with stones or seeds. Dowel pushed into openings and sealed for handles.
- Scrapers – children chose pieces of wood or cane and cut nicks in them. One chose corrugated cardboard and mounted it on a piece of wood.
- Pan pipes – tubes of cardboard cut to graded lengths and mounted in various ways onto card

or board with card, decorated. One interesting one was mounted on a detergent bottle.

I have selected some of the written work of the children and the main feeling I get from looking at this is that the children had to be much more precise and systematic in their writing than they have previously had to be. Samples of written work can be found in the Class 2K Book, *Musical Instruments*.

6 The display

The display in the classroom gradually built up as the instruments were completed. However, we had other aspects of the topic to explore. The various agencies of sound could be investigated and so paintings of mechanical objects which can produce sound – such as aeroplanes – were contrasted or compared to live objects which produced sound such as a person shouting, clapping, etc. Good paintings were produced and displayed. Later on we made up or thought of sound 'words' and used letter templates to cut out letter shapes and made sound words to hang up from the ceiling.

One child brought in a musical box (after the visit to the Mechanical Music Collection) and after observing its workings we made an attractive display of musical boxes in the classroom for a short time.

I drew a large diagram of the ear and its internal features and the children coloured and labelled it; this also formed part of the classroom display.

The instruments were always a focus of interest and constantly handled and played by the children when they came in after play or at other times when they were free to play; the flowerpot chimes were the only instruments subject to damage and two of them were broken.

The doorbell chimes brought in by a parent were also hung on the back wall and made a lovely sound showing the size of the tube made a difference to pitch.

The cloakroom was decorated with work about the trip to West Dean College and the Mechanical Music Collection.

Eventually the best parts of the display were

moved to the school entrance hall and stayed there until the end of term.

Photographs of these displays can be seen in the additional book by Class 2K, *Musical Instruments*.

7 The Visits

In the fifth week of the term we made a visit to the West Dean College where courses in early musical instruments take place. The main part of their work is the making of violas and the Director of the department very kindly showed the children the various aspects of instrument making starting with the types of materials used and the procedures gone through to achieve the very beautiful looking instruments on display. We were actually in a workshop with tools and materials easily observed. Many questions were asked by children and accompanying parents and I was surprised at the detail recalled when writing was done over the following few days. We stayed here for about an hour and after lunch visited the Mechanical Music Collection near Chichester. This too was an enlightening experience as well as entertaining, as many instruments were made to play for us.

The flavour of these two visits can be appreciated better by reading the children's writing and looking at the pictures they made. These can be found in the book, *Musical Instruments* by Class 2K.

8 Cross-curricular applications

Science

The exploration of sound, this topic fulfils AT14 National Curriculum.

Level 1   Pupils should: know what sounds can be made in a variety of ways.
Level 2   Know what sounds are heard when the sound reaches the ear.
Be able to explain how musical sounds are produced in simple musical instruments.
Level 3   Know what sounds are produced by vibrating objects and can travel through different materials.
Be able to give a simple explanation of the way in which sound is generated and can travel through different materials.

The topic also goes some way towards AT1 – the general approach to exploration and science. Also, AT2 – The Variety of Life, and AT3 – Processes of Life.

Maths

Measuring using standard measures, simple ideas of ratio (the pitching of size). Counting and sorting. Estimating before making. Simple concepts of volume, capacity and length.

English

The discussions, writing and reporting. The poems, stories, sound words. Vocabulary – words used to name parts of musical instruments, words used to describe sounds.

History and Geography

Very briefly – the early musical instruments and pictures of people and the costumes they wore. Some discussion of places from whence wood to make violas and bows came.

Technology

- AT1 Identifying Needs and Opportunities
  The need to make musical instruments and to produce musical sound for enjoyment. Level 1 and 2.
- AT2 Generating a Design Proposal
  To produce a realistic, appropriate and achievable design.
  Level 1 and Level 2
  The designs were from a choice of pre-designed instruments but many modifications were achieved.
- AT3 Working to a Plan – Make an Artefact, System or Environment
  Level 1, Level 2 and Level 3
  I see this area as the strongest in this topic and also most relevant to the age group.

Appraising

- AT4 – Pupils should be able to develop, communicate and act constructively upon an appraisal of the process, outcomes and effects of their own design – also the design from other times and cultures.
Level 1, Level 2 and Level 3
These aspects of the topic are also strongly featured.

The weakest feature of the work, from a design point of view was that children may not have been given sufficient opportunity to design their own instrument and I see this as a challenge for the future when I shall be doing this topic again as part of the Science scheme of rolling topics related to attainment targets right through the primary age range in the school.

Music

Appreciation of instruments – pitch and why instruments look the way they do. Unfortunately, we did not get time to do much music making on the instruments, as time just disappeared. However, there will be time for that in the future and I hope to make some recordings. We did use a computer program about pitch very successfully.

Art and Craft

- The instruments were made to look aesthetically pleasing. Instruments have to be seen when being played before a live audience. Paintings of sound-producing things.
- Drawings of instruments.
- Illustrations on *Noisy Poems*.

PE and drama

Using the body to produce sound and rhythm. Listening and moving to mechanical, electronic and natural sounds, for example, thunderstorms, rain, windy weather, etc.

RE

Briefly discussed musical instruments used in religious festivals or of different cultures.

9 The Assembly

After the half-term break we had three days to prepare an Assembly for the Thursday morning when parents were invited. I should have liked longer to prepare the children as they seemed to lack confidence on the day of the Assembly and I did not feel it did justice to all the hard work they had put in.

However, here is an outline of the Assembly:

1 Hymn – 'He Made Me'.
2 Sound words marched silently across stage.
3 Talking about sound and hearing with large diagram of ear to help.
4 Talking about the sound tins game with graph to show people's guesses.
5 Reading out sound poems composed by themselves.
6 Showing pictures and talking about the visits.
7 Showing the instruments and telling how they were made.
8 Telling story of Joshua and Battle of Jericho.
9 Reading from *Noisy Poems*.
10 Reading of child's own prayer about the gift of hearing.
11 Final hymn – 'I Listen'.

Appendix

School resources

Lively Craft Cards set 2 – *Making Musical Instruments*, Mills and Boon.
Make Music and Making more Music, Richard Addison, Holmes McDougall.
The Sense of Hearing – Science 5–13 – Stages 1 and 2, Macdonald.

Schools Library Service

Buttons, Linda Yeatman (Hearing Dogs for the Deaf).

Making Musical Instruments, Margaret McLean, Macmillan.

Making Musical Sounds, Mary Southworth, Studio Vista.

Experimenting with Sound, Alan Ward, Dryad.

Musical Instruments in Colour, John Gammond, Blandford Press.

Your Nose and Ears, Joan Iveson-Iveson, Wayland.

What Happens When you Listen? Joy Richardson, Hamish Hamilton.

Hearing, Mary Gribbib, Macdonald.

Music Maker, Robina Beckles Willson, Viking Kestrel.

Scrape Rattle and Blow, Chris Desphonde, A. & C. Black.

Hearing, edited Rex Catherall, Wayland.

The Pied Piper of Hamelin, Robert Browning, Spencer.

Alarm Bells, John Escott, Hamish Hamilton.

Noisy Poems, Jill Bennett, Oxford University Press.

DEVELOPING A STRUCTURED APPROACH TO DESIGN AND TECHNOLOGY AT KEY STAGE 2 AT MARSTON GREEN JUNIOR SCHOOL
Reported by John Dainty

Marston Green Junior School has around 200 pupils aged 7 to 11, roughly half from the nearby Chelmsley Wood Council Estate. There are eight classes, with two parallel classes in each year. Two years ago the school adopted a whole school thematic approach with a 12-term topic plan integrating most aspects of the curriculum. With the help of the LEA guidance document and using the organisational planning strategy shown in Fig. 7.1, topic webs were drawn up by each year team. With the confidence and knowledge gained from in-service training, opportunities for the Design and Technology activities were identified.

The second stage included mapping chosen activities with the four aspects of the programme of study for Key Stage 2 and allied attainment targets. Continuity of planning was ensured by using a planning sheet developed by the LEA and shown in Fig. 7.2.

During the Summer term the planned activities were trailed as part of the normal school curriculum, and staff meetings set aside for discussion of progress.

Some aspects of Design and Technology were not being covered and would need attention in

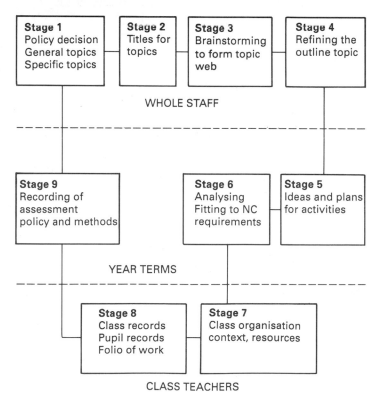

Stage 1 Policy decision General topics Specific topics	**Stage 2** Titles for topics	**Stage 3** Brainstorming to form topic web	**Stage 4** Refining the outline topic

WHOLE STAFF

Stage 9 Recording of assessment policy and methods	**Stage 6** Analysing Fitting to NC requirements	**Stage 5** Ideas and plans for activities

YEAR TERMS

Stage 8 Class records Pupil records Folio of work	**Stage 7** Class organisation context, resources

CLASS TEACHERS

Fig. 7.1 Organisational planning strategy

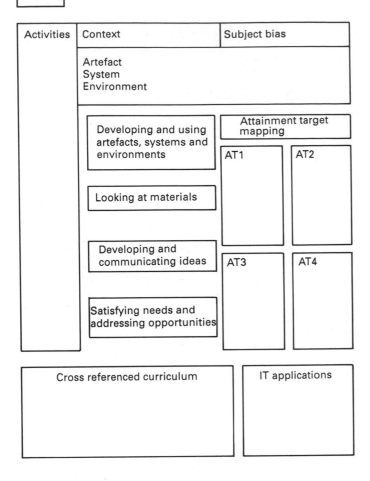

Fig. 7.2 Design and Technology planning sheet

future planning. Some of the problems identified included:

- Wider media use, especially in clay, food and textile areas needed; greater attention to Attainment Target 1
- The need to develop appropriate methods of recording individual pupil achievement
- A higher profile needed for the business context
- The need for a more balanced approach, extending the scientific route to include more creative and aesthetic elements

The work of Year 3 children revolved around the theme of 'Water'. Two Design and Technology projects were undertaken, the first based upon the design and development of an Environmental Studies Area at the school, including a marsh and pond. The local power station at Hams Hall kindly provided help in the form of advice and some plants. Sadly, vandalism was a big problem, but the Year 3 pupils persevered and took a positive approach to the problem by working on ideas to counteract it.

The second project involved the design and manufacture of an ice-lolly, and the design of its packaging. The children carried out consumer research in school and at the local shops, so setting it in a business context.

Year 4 children worked on 'Buildings', investigating the materials builders use, damp-coursing, foundations, the bonding and making of bricks, and so on. They created structures from the information they had gained and tested them for strength.

Fig 7.3 An example of a completed Design and Technology planning sheet

ACTIVITIES

STAGE 1
Short investigations using lego technic – pulleys, cogs

Safe and proper use of tools

Introduction control software

STAGE 2 Structured design brief the 'Black Box'
Make an axle turn by turning another at right angles to it

STAGE 3 Main design projects
Choice from

1 FROM VISIT TO IRONBRIDGE
(a) Inclined plane
(b) Crane
(c) Blowing engine
(d) Forge hammer
(e) The Ironbridge
(f) Coal mine

2 MODERN HISTORY
(a) Robot
(b) Controlled enrivonment
(c) Automatic traffic control
(d) Automatic sorting machine

CROSS REFERENCED CURRICULUM

Science
Attainment targets 10 to 13 – forces and energy.
AT11 – electricity, 12 – IT and 6 materials

Mathematics – measurement and 3D shapes – nets of solids

Language
Reading for information
Industrial revolution booklet – research and empathy writing

CONTEXT INDUSTRY TOPIC

ARTEFACT (s) Series of models
SYSTEM Mechanical systems used in Victorian machinery. Modern electronic systems
ENVIRONMENTS Difference between industrial revolution conditions and modern factories

Developing and using artefacts, systems and environments

Use video/books to research industrial revolution
Use this information and knowledge gained through lego work to design drive systems. Organise themselves and allocate time within specified timescale.
Test and modify initial drive and control systems

Looking at materials

Instruction in safe use of following – shaper saw, hacksaw, bench hook, drill, glue gun, craft knife, soldering iron, wire stripper.
Select materials from – 8 mm timber multiboard, dowel, card, softwood offcuts PVA and hot glue.
Stress limited resource – avoid waste.

Developing and communicating ideas

Provide plans of models showing detail of drive and switching systems.
Keep a record of development and changes made.

Satisfying needs and addressing opportunities

Recognise through modelling the importance of the right working environment and how that environment might be improved.
Recognise the need for safety measures in the work place.

SUBJECT BIAS Science and CDT

ATTAINMENT TARGET MAPPING

AT1
4c discuss what it must have been like to work during industrial revolution
4f Research factors which led to development of iron industry at Ironbridge.
Discuss need for pleasant and safe working environment

AT2
4a Record development of systems from first 2 cog experiment to final model
4b Development of building in which to house model
4c Estimate and plan use of resources before first cut.

AT3
4b apportion task within group
4e draw diagrams to modelling

AT4
Review continually the development of models.
Possible transcripts of recordings of group discussion.
Illustrate to other groups implications of factory environments by use of models.

IT APPLICATIONS
4a Use of DTP package impression to produce a record of work in the form of a Victorian newspaper
4b Use of 'BITS' to control movement and sense environment.
4c Use of 'Beasty' to control movement.

'Feed the World' was the topic for Year 5, considering the problem of raising water for irrigation using wind power. A well was made from a large sweet jar and, using the principles of wheels and cogs, the pupils designed and built windmills to raise water from it. They also designed a farm to illustrate how all its elements fitted together to create the best possible environment for the farmer.

Year 6 tackled 'Industry', comparing the environments and systems of Victorian times and the modern age. The work was particularly challenging because there was only one term in which to give the children all the skills that children would, in future, progressively acquire throughout Key Stage 2.

The children were thus equipped in three stages (Fig. 7.3). Stage 1 dealt with the necessary knowledge, skills and confidence. Children were taught how to use tools such as hacksaws, drills, glue guns, bench hooks, soldering irons, etc. safely; how to measure accurately so that model designs fitted exactly. Lego Technic was used to demonstrate belt drives, pulleys and cog wheels. The children were also shown how to use computer control software.

Next they applied knowledge and skills to a structured design brief. They were given a shoe box with two pieces of dowelling sticking in two of the sides, at right angles to one another and they had to find a means of making one dowel turn when the other was turned.

Stage 3 provided a choice of projects. Half the group studied the Victorian industrial scene and the other half industry today. The first group produced models including a furnace blowing engine, a steam-driven forge hammer and a coal mine. The second group's products included a robot which picked up articles, a model factory and an automatic sorting machine.

The children were able to use their computer experience to control lights, motors and sensing devices so imbuing it with a degree of realism.

These pilot projects highlighted some problems in this new high resource area; resources are in short supply. Another problem is time. Although Design and Technology is an exciting curriculum area, it is time consuming. Our efforts have brought this home to the staff, reinforcing the need to plan and record systematically as a whole school.

There are many positive results, however. Staff have found it a valuable awareness-creating exercise and it has built up their confidence. Their ability to recognise design opportunities has been greatly enhanced. The initiatives have revealed the importance of structure and of a progressive approach to the four years of Key Stage 2.

Overall, staff and children gained considerable confidence and experience and the school has made tremendous progress in working towards the implementation of National Curriculum Design and Technology, especially with regard to planning and the organisation and management of resources and the working environment.

THE KEY STAGE 3 TECHNOLOGY, RESEARCH, INFORMATION, PROBLEM SOLVING COURSE (TRIPS) AT GREENHEAD
Reported by G. W. Asquith and D. Smith

Greenhead Grammar School, Keighley, West Yorkshire, is a 13–18 mixed comprehensive school of about 1000 students. Children progress to Greenhead through a middle school system in the areas to the north and east of Keighley town centre.

Design Technology is housed in nine purpose-built rooms of various vintages. Information Technology is accommodated in three. All 12 rooms have been refurbished using a self-help approach and a small financial contribution from TVEI. Our teaching of Technology needed developing so that it was no longer seen as a separate entity, but rather as an integral part of our curriculum. The Technology Research Information Problem Solving (TRIPS) course commenced four years ago and is still developing.

Word processing results of a survey

Students research for electronic items and costing

TRIPS

A cross-curricular Technology course

This is designed to cut across curriculum boundaries, and is based on Electronics, Design Technology, Information Technology, Food and Fabric Technology and Media Technology. The course aims to show that Technology is not a subject with a particular content but rather a process which develops skills throughout the whole curriculum. The TRIPS course aims to provide students with skills to approach problems through investigation and research, to acquire knowledge to solve relevant and real technological problems, to foster initiative and resourcefulness, to be technologically aware, to be able to co-operate and accept responsibility in a group situation, to improve communication skills and to be able to make judgements which are aesthetically, technically, economically and morally sound.

The course has two elements:

1 An introduction – to raise awareness of the scope and range of opportunities in each of the five contributory areas.

	Business tech	Media tech	Electronic tech	Design tech	Food and material tech
1	Personal database	Petrol promotion	Security system	Company name and logo	Packed lunches
	Company payroll	Newspaper advert	Warning device	Planning of site	Ergonomics
2	Ergonomics databook	Menu cards	Electronic sign	Company transport	Company uniform
	Product costings	ID cards	Cash/EFTPOS terminal	Planning the building	Protective clothing
3	Employing staff	TV/radio advert and jingle	Intercom system	Eating unit	Interior design
	Promotional booklet	Promotional gifts	Keeping hot/cool	Playground	Presentation buffet

Phase (vertical label on left, beside rows 1, 2, 3)

Fig. 7.4 TRIPS Small Business Scheme: stage/work matrix

Making an electronic sign

2 A Small Business Scheme – the major element of the course, divided into three phases each containing different tasks in each phase (see Fig. 7.4). Tasks may be directed at internal company management or towards the current project and allow pupils to tackle problems encountered by a small business in the marketplace. Groups of pupils are responsible for developing various aspects of the company and are expected to produce and present finished project work in a set format including, where appropriate, a finished piece of technological hardware. Pupils are assessed on a regular basis, both by verbal appraisal and by a written personal Record of Achievement (see Figs 7.5a, b).

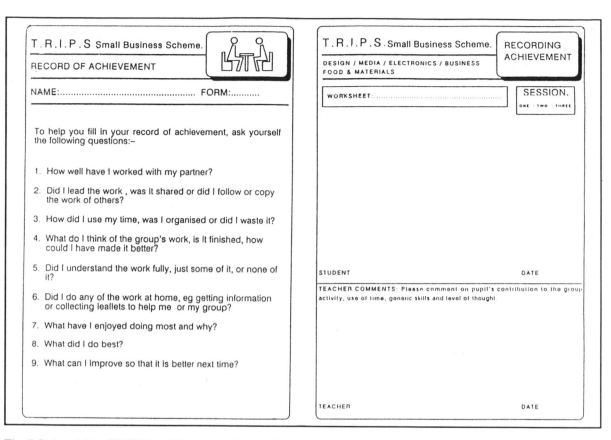

Fig. 7.5(a) and (b)   TRIPS Small Business Scheme: Record of Achievement

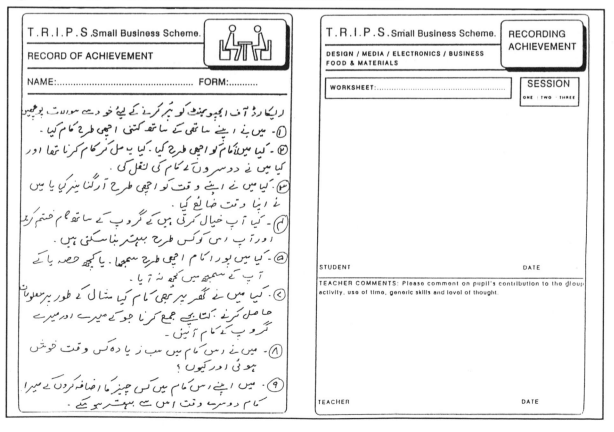

A typical scenario for the TRIPS Small Business Scheme would be a small design company, The Design House (Fig. 7.6) which has been developing over the last two years and has now been asked to advise on the production of a total working package and corporate image to suit a client's needs. The company has been called in to advise a national catering company with outlets in motorway service stations, which needs to remain competitive and retain its high profile in the market place. The Design House has been contracted to produce a complete design package encompassing company name, logo, transport and uniform, interior design, furniture design, promotional materials, provision of children's play area, communication and security.

The structure has proved to be crucial to the perceived success of the scheme. In a typical year group (and at present we teach TRIPS only in Year 9, due to pressure to teach to an examination in Years 10 and 11), there would be ten registration forms. Each of these would have 20–24 children of mixed ability, race and gender. The TRIPS scheme would be delivered to half the year group at a time (six form groups). Each of these forms would be randomly allocated to a Design and Technology teacher. There is also an additional teacher, who provides support for students where necessary. The teacher and form would work together to form a 'company'. The company's first task would be to decide the responsibilities each student would undertake within each area of the company and who would take on the role of manager in each of the small groups of students (three or four) going to each area. The managers' main tasks are to organise the work in each area (Electronics, Design Technology, Information Technology, Food and Fabric Technology and Media Technology Science) to meet the demands of the company and to report back to the management meeting on progress and future targets at the first lesson each week. TRIPS is allocated 3×50 minute periods per week in a 30-period week.

Practical application

It is important to understand that TRIPS is a small business simulation, not a mini-enterprise scheme. The major difference is that students are involved in a much larger range of problems and situations than a small company is likely to face in the marketplace. We have attempted to make the company, its operation and the work it is involved in, as realistic as possible within a school situation. There are some aspects of TRIPS which are not found in any other course.

Each of the 12 companies (forms) in the year group work as the small design company, The Design House (Fig. 7.6) using the same brief but reacting to it and producing results in very different ways. The tasks are fully resourced and have an accompanying worksheet designed as a guide to enable all students to achieve success in each one.

The 36 activities are organised into three phases so that a natural progression of work and information can flow through the scheme. Some tasks need to be done first, since others depend on them. The structuring of the work has allowed a number of study paths to be developed within the tasks, so that students can follow a theme throughout the scheme. For example, students who have worked on the Ergonomics task in Food and Materials Technology may follow this through into Ergonomics databook (Business Technology Phase 2) and then into Children's Playground (Design Technology Phase 3) (see Fig. 7.4). Students are allowed to follow a study path or make a company choice about where they will move at the end of each phase.

The three phases are treated as separate stages, so that we can achieve a number of tasks. First, it allows all students to experience three of the six areas, which may be choices based on personal interests or company-directed as a result of the management meetings. Secondly, it allows a presentation and evaluation session for each company at the end of each phase, in order that current progress and work can be reviewed and enables Recording of Achievements to occur. At the end of the whole scheme each student will have three Record of Achievement profiles, one for each area in which they have worked. Finally, each company gives a presentation involving the display of students' work and a verbal explanation of their design proposals to all the other companies.

Fig. 7.6 TRIPS Small Business Scheme: The Design House

BACKGROUND

The company is a small firm formed 5 years ago by two colleagues. They initially did small contract design work, producing such items as advertising materials, company logos, hand-books etc. The company has developed during the last two years in a number of areas due to demand. They needed to be able to produce a total design package which covers a wide range of extra areas, such as ideas for fast food, retail outlet design, stationery, packaging work, clothing and uniform. The demand continues to be for one of total working package and corporate image to suit a company's need.

The firm has just brought in people who are experts in video and computer graphics. A studie has been set up and they are now able to produce advertising/display materials using their two new Hi-tec tools.

The firm now has experts in the areas of:

Electronics, Graphic Design, the Media, Food preparation and presentation, Business organisation and Fabrics.

SITUATION

A national food company with outlets involved in motorway service stations needs to remain competitive in this ever increasing market.

The Design House has been asked to look at the following areas:

- Company name and logo
- Company transport and livery
- Exterior layout of a sample site
- Company uniform and protective clothing
- The interior design
- Provision of a play area for children
- Furniture design
- Promotional booklet
- Expansion plans
- Signage
- Communication on site
- Security
- Menu cards
- Promotion in the media

Cutting out card patterns for waiter/waitress head gear

There is still much to do, but we feel we have a structure that will take us through and meet National Curriculum requirements at Key Stage 3. The TRIPS course has injected new life into the department. It has provided remarkable motivation for our students. Our discipline problems are minimal and low achievers have found a medium through which they can shine amongst their peers and form bonds of trust. However, the demands on teachers are greater than ever, as we are now not only educators but have also to take on roles as managers and facilitators.

We are most gratified that other schools have also found the work of value and as a team we have been very busy teaching teachers from various areas of the country. Future developments are in the pipeline to include three more curriculum areas – those of Languages, Humanities and Expressive Arts.

Appendix

Photocopiable Teaching Packs are available for this course at a cost of £26.50 (cheques payable to TRIPS) from Brian Smith, 15 Willow Tree Close, Longlee, Keighley, West Yorkshire BD21 4RZ. Tel. 0535 664819.

PUTTING THEORY INTO PRACTICE IN KEY STAGE 3 AT VANDYKE UPPER SCHOOL
Reported by Peter Jones

Vandyke is a 13–18 Upper school in a small country town on the western edge of Bedfordshire. Its intake is truly Comprehensive, with students coming in from the surrounding villages, as well as from the town. We felt it essential to face the challenge of National Curriculum Design and Technology, earlier rather than later to give time to iron out difficulties and develop a course which not only reflects the spirit of Design and Technology but also includes the rigour needed to develop a real technological capability. It also gives us time to integrate the programmes of study rather than structuring a course to which elements have to be tacked on.

With the encouragement of the Headteacher, the Design and Technology area has been restructured, both physically and by the creation of a new faculty in Design and Technology. Refurbishment is a key element to the subject's development – the need to project a well-organised, warm and inviting environment to encourage a positive response from students, parents and other members of staff. The faculty is made up from CDT and Home Economics, with Business Studies and Art affiliated. This is because Business Studies is heavily involved with the Community college and Art with Performing Arts. Information Technology is regarded as a cross-curricular issue and so IT facilities are being built into the refurbishment and the course of study.

Year 9   Design and Technology course – structure and content

Term 1 (Autumn term), 14 weeks – Festivals and Celebrations

Approach

In the first term we decided to pick a theme first and ask students to identify their own contexts from it – a very different approach from where the context comes before the theme. The theme chosen for the first term was Festivals and Celebrations and it was taught through an open-task and integrated-material approach. The initial purpose was to see how the students coped with designing and making using appropriate materials, being process led rather than material led. Coming from a wide range of middle schools, the students' previous experiences were varied and many had been taught Design and Technology through separate disciplines. So our initial objectives were, firstly, to offer a holistic experience in Design and Technology, and secondly, to encourage a team approach.

The introduction

At the start of the course, students were shown the faculty as well as any new or additional facilities which their middle schools would probably not have had. This was followed by a video and slide presentation in the school theatre to introduce the theme and help trigger ideas.

In the homebases, the groups would start with a brainstorming session on Festivals and Celebrations, with discussions on their importance and how they relate to us. This was followed by a brainstorm on one specific festival/event with possible ideas on a creative approach. We did not lay down any rules, only that we would like students to experience all the component areas

within Design and Technology. This section took a long time and a lot of hard work. We tried to encourage the idea that the work be interrelated, i.e. if they were preparing a toddlers' party that the party food, bags and tableware should have the same central idea. The photo of the Teddy bears' party gave a good idea of this.

The lessons

Each student was assigned to a homebase with an adviser to monitor progress. They met for 15 minutes at the start and end of a two-hour block. In the middle section the students would book into a specialist area of their choice, to develop and realise their ideas. Over the course we expected students to work through the Student Management Process shown (Fig. 7.7).

The General Lesson Plan (Fig. 7.8) shows the overall plan which would be altered when the situation required it. There were three main categories of people involved in this teaching experience – the adviser, the key specialist and the support staff. The addition of the support staff allowed flexibility among these roles and allowed

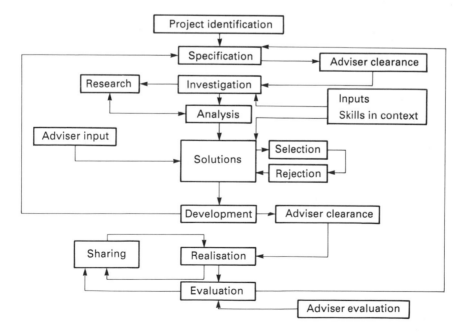

Fig. 7.7 Student management process

	ADVISER	STUDENT	SUPPORT
BRIEFING	Take register Give input if necessary Check H/W Check students' progress/work	Check diaries/red folder Fill in planner Discuss in team way forward, ensure group organised	Check all rooms in case need for help
ACTIVITY	Key specialists give inputs when opportunities arise (try to support these with hand-outs)	Go to booked area Note down any staff inputs/ demonstrations Go to book rooms for next week	
DEBRIEFING	Check register Check students' progress Check H/W set	Write up planner Plan for next week Fill in adviser sheets of where you will be working next week, identify H/W	

Fig. 7.8   General lesson plan

movement from week to week. Only the adviser was permanent to ensure continuity for the students.

We decided to clearly define each role to help everyone to understand the interrelationship of the team.

The role of the Design and Technology advisers

The five main roles for the advisers are:

1 *Explaining:*
  - the aim of the course
  - the facilities available
  - the safety rules
2 *Developmental:*
  - developing students' understanding of the holistic nature of the design process and the contributions of all five areas
  - developing good team work
  - developing good practice
3 *Monitoring:*
  - general progress

  - ensuring students keep diary/records up to date
  - making/collating assessments
  - marking tests (if any)
  - writing reports
4 *Feedback:*
  - to planning group about organisational problems, etc.
5 *Guidance:*
  - about GCSE and other Key Stage 3 courses in Design and Technology.

The role of the support staff

The support teachers were required to undertake between them a number of roles and to allow flexibility within the system:

- Helping to identify and overcome organisational difficulties.
- Providing support for staff operating in new areas.
- Encouraging staff to take a broader view of Design and Technology.

- Ensuring that students' experience of Design and Technology is not restricted by the subject specialism of their adviser.
- Monitoring the progress of the course and reporting to the planning meeting group.
- Helping with evaluation tasks on behalf of the planning group.
- Relieving advisers in order to enable them to undertake wider functions.

The role of the key specialist

This is the third member of the team whose role is to offer specialist advice in a number of ways:

- To give students an overview, through discussion of the whole project.
- To have general Design and Technology skills to ensure a common approach throughout the faculty. This is to ensure a common, agreed level in quality control and expectation.
- To offer subject-specific skills and to ensure rigour and quality in the development of the students' ideas.
- To support staff who wish to call upon their individual key specialism.

The facilities

The areas available were usually two or three Design and Technology studios, which are multi-material, a Home Economics room, with cookery and textile facilities, and an Art room, with three-dimensional clay and printing facilities. Resource rooms with IT facilities are being developed and are central to both the course and the faculty's layout.

The work

Having worked through this once, it has become clear not only how to improve the general management, but also how to monitor the students' progress. The main difficulty was to encourage students to think for themselves and not to expect us to tell them what to make. For some students this freedom was hard. The photos show some of the real successes, where students have produced items reflecting not only their theme, but also the real spirit of Design and Technology.

The evaluation

Full staff and student evaluations took place at the end of the term and part of this was the presentation of the students' work. It was generally agreed that the open-task approach was a good way to start. Students enjoyed the freedom to choose as well as being able to work in a range of areas, as and when needed. The staff enjoyed the wide range of topics which came forward and found it a stimulating and creative experience. However, we did have problems with students getting access to materials and equipment quickly and efficiently; and sometimes the demands on staff, due to the wide range of activities, were almost excessive.

Term 2 (Spring term), 10 weeks – Restaurants and Catering

This term is structured very differently. We are undertaking a closed project which is heavily teacher-directed. The purpose of this is to build upon those areas of concern identified by the student and staff evaluations. The work is individually based and tightly structured to ensure that these areas are tackled. This has been clearly explained at the beginning of the project and included in the student resource notes.

The seven main areas of concern are

1 Planning
2 Researching, selecting and sorting this material
3 Communication ideas
4 Practical skills
5 Clearly identifying needs
6 Teamwork
7 Evaluation

The task for the students this term is to construct a business plan, going through stage by stage, proposing to open a new pizza restaurant in the town centre. The best will be selected and presented to a local bank to be judged. The bank will be sponsoring an award to be given to the winning student.

The theme of birth is a perfect example showing the complete spectrum of Design and Technology

Masks, Monster Biscuits and a Candle Holder, part of a planned Hallowe'en party

A Chinese Boat made in a range of rooms with a range of materials

A Ghost Cake showing the variety and humour in food

A Troll Mask made of Mod-Roc

A children's party held in the school's crèche. The theme of the Teddy bear runs through, from the large wooden Teddy at the back which holds cards, to the printed cloth and party hats

A Candle Stick, Mask and Banner celebrating the Chinese New Year

We have constructed nine sections that each student must complete.

1 A study of the local cafes, restaurants and take-aways
2 Research on pizzas and healthy foods
3 A typical meal
4 Interior decoration
5 The menu
6 Financing the new pizza restaurant
7 Good team work
8 The advertising campaign
9 Evaluation – maintaining our quality standards

The practical element was concentrated on Home Economics, Graphics and Design skills; we felt that to add to the CDT element within this business-plan approach, would be out of place and not appropriate.

Term 3 (Summer term), 12 weeks – Conservation

The planning for this term will be similar to that of the first term. The theme of Conservation was chosen, not only because it is topical, but also because it will link with the Geography department, who will be running a related project at the same time. Group work will be optional, but the open task approach will be repeated.

What has been learned about the process of change?

What advice could we give to someone who is about to embark on such a curriculum reform? Below are some of the main issues that we have identified.

1 Quick curriculum audit of the area in relation to the National Curriculum requirements. Do not take days, and keep it simple.
2 Ensure the management team are involved in details, not just a brief encounter in the corridor or half-hour chat. Ensure that the LEA staff are also involved, they add weight and give a

broader understanding of the issues. Some of the management issues are staffing, building improvement, development fund, capitation, INSET for staff to gain new skills and teaching methods, as well as time for planning and preparation.
3 Take time in planning and involve *all* in the discussions, it ensures ownership and it builds a faculty identity.
4 Plan Key Stage 3 as a whole. For most secondary schools this is easy, but for the Middle and Upper schools this presents a range of problems; liaising has to be done after school.
5 Anxieties. At present there are lots of anxieties within schools and between schools. Each school must identify its own and find its own way through them. Planning groups and faculty meetings helped.

The future

This development is only at the beginning and we have a long way to go. Major issues, such as assessment, continuity and ensuring rigour and progression are still to be addressed, but concentrating on the spirit of Design and Technology and not being tied by the fine print at this stage gives the students a far more coherent and enjoyable experience. It also gives us time to develop new approaches and methods of working.

Acknowledgements

The author thanks all staff involved but especially John Duxbury for helping with the background detail, Terry White for supporting the developments, and Linda Moore for holding everything together.

Reference

Lever, C. (1990) *National Curriculum Design and Technology Key Stages 1–3*. Stoke-on-Trent: Trentham Books.

CHAPTER 8

Conclusion

This concluding chapter reviews some of the ways in which Design and Technology may continue the progress that has been reported in previous chapters, notably through the development of quality, and some of the prospects that may be realised.

The title of this chapter is really a misnomer. There can be no conclusion to a book on teaching Design and Technology. The subject area and the ways in which it is organised and taught are constantly changing and developing. It has changed dramatically as this book has been written and will continue to change – with possibly even greater speed. The speed is essential if Design and Technology in school is to keep pace with Design and Technology in society at large. And if it is to lead and guide society – as many wish it to do – then Design and Technology must develop even more rapidly still.

Even though we are in rapid motion, however, there are important points for us to notice if we are to continue our journey and to control its direction. Above all we must realise that the success of our activities is not assured; we must work constantly to try to achieve it. The inclusion of Design and Technology in the National Curriculum of England and Wales or any other country does not guarantee it a permanent place or even credibility or status. To ensure enduring respect and recognition the quality of the product is crucial. Andrew Breckon, a member of the Working Group on Design and Technology, noted this early. Breckon (1990) writes:

The curriculum area of Design and Technology must use strategies to develop pupils' capability to investigate, design, make and appraise, while contributing to pupils' economic and careers awareness, stimulating originality, encouraging enterprise and emphasising quality.

To realise the full significance of this need requires us to recognise the new meaning of quality. The UK, like most manufacturing countries, has always prided itself on being able to make quality products. But 'total quality' has been more elusive. We have made machines of immaculate precision – but difficult to maintain or ungainly in design. We have made elegant motor cars but with unreliable components and a propensity to rust. We have made superb tableware – but with an embarrassingly high percentage of rejects of third and even fourth quality.

It has been left to German and – more spectacularly – Japanese and Far East industry to demonstrate the concept and the reality of total quality – throughout all aspects of the product and throughout the production line. The techniques whereby this is achieved through quality circles, quality control, contracts, workplace conferences and much else, are now being adopted by industry in all countries. The spread of interest in quality standards in Britain and in the British Standards BS 6057, and the concern surrounding the implications of the 'full' Economic Community in

Europe, are evidence of this awareness. It is essential that similar awareness be aroused in the schools if they are to deliver a quality product to all their clients – the pupils and their parents and employers. And the most appropriate place from which this initiative may spring is surely the Technology Department.

To achieve consistent quality of Design and Technology experience has been an underlying theme throughout this book. It requires a keen awareness of our starting points and the nature of our knowledge base. It involves a perceptive knowledge of curriculum opportunities and the legislative context in which they occur. It needs a capability to assess creatively; to use assessment as a personal professional tool to ensure quality control rather than simply to satisfy imposed regulations by Government, examining board or school administration. Ensuring effective availability to all pupils and providing competent, sensitive and efficient management are also crucial – as relevant chapters have demonstrated.

It goes without saying that the quality of the teacher is a central element of total quality. Fortunately, this is widely recognised – for instance in the 'Schools Charter' developed by *The Independent* newspaper (1991) which states unequivocally that 'the most important single factor in raising standards is the quality of teachers'.

Having delivered a quality product, it is vital that this is recognised widely and taken into account by parents, employers, pupils and politicians. And this recognition must be achieved without distorting the very nature of the product as David Layton, also a member of the Working Group on Design and Technology, noted in an unpublished lecture at Warwick University in 1991:

> At the end of the day, assessment procedures have to win acceptability from a number of different social groups, each of which may employ different criteria. Delicate judgements are required to ensure that the effects on teaching and learning of Design and Technology are not baleful whilst, at the same time, the practical, economic and political credibilities of the results of assessment are retained.

If total quality and its widespread recognition can be achieved then the transmogrification of Design and Technology will have achieved far more than status and security in the curriculum. It can have a general effect on the curriculum as a whole as Roberts (1990) notes:

> The creation of a technology curriculum for all could radically change general education. The damaging partiality that has always been inherent in the popular conception of the sufficiency of the 3Rs or the 'basic skills' might be removed.

Roberts's comments bring us back to the fundamental issue that dominated the early chapters of this book – the persistent divide between the 'academic' and the 'practical' subjects in the curriculum and the parallel gap between managers and makers in industry. It is a status gap which has dominated education in many countries – notably the UK – and has had immensely damaging effects on both industry and social structure.

Fortunately, there are signs that the problem is now being recognised. The White Paper *Education and Training for the 21st Century* (1991) subsequently endorsed by the Prime Minister in January 1992, offers a clear analysis of the problem and proposes a restructuring of institutions and examinations to attempt to close the gap – or even eradicate the divide. The proposals of the White Paper include:

> An Advanced Diploma which would be awarded to students who gained two 'A'-Levels at grade C or above – or equivalent vocational qualifications, or a combination of the two – with 'AS'-Levels. Such new National Vocational Qualifications (NVQs) would provide 'a clear ladder of progress'. They would be awarded at levels rising from one to five – level one being rudimentary competence in work-related skills, with level five being competence in very complex and wide-ranging skills.
>
> The new structure of qualification will 'offer the prospect of a workforce with first class skills to produce the wealth on which our society depends for its standard of living'.

The White Paper was accompanied by new instructions to the National Council for Vocational Qualifications (1991):

Many young people want to keep their options open. They want to study for qualifications which prepare them for employment in a range of related occupations and keep open the possibility of going into higher education. Adults who wish to prepare the ground for a major career change also want access to broader, more general vocational qualification.

The aim should be to have the first general NVQs at levels 2 and 3 accredited in time to be available in colleges and schools from September 1992. Working closely with the three main awarding bodies, City and Guilds, Business and Technical Education Council and Royal Society of Arts, it should be possible to make rapid progress towards modifying some existing qualifications to bring them into line with the new criteria for general NVQs very quickly, and accrediting them.

The coincidence of this new initiative with the introduction of Technology into the National Curriculum is one of the most promising prospects of education for the 1990s and beyond. Cynics will point out that there have been promising initiatives before and that the 'dual system' of academic and vocational qualifications still remains throughout the White Paper. But to achieve parity is perhaps the essential prelude to the eradication of the divide.

Yet an optimistic overview of the future should not obscure the magnitude and difficulty faced by teachers of Design and Technology who, as we have seen, have to combat not only the entrenched attitudes of school and society but also to fight for teaching time, resources and recognition.

As always there are temptations to take short cuts. Any opportunities for variation from Technology at Key Stage 4 may tempt some schools to distance some, most, or even all of their pupils from the subject and so avoid the complexity and cost of delivery of Technology at Key Stage 4. In so doing they would deny their pupils not only the understandings and experience of technology but also the recognition of the proper nature, status and importance of the subject in relation to others. It may even be that some schools will rush some of their Key Stage 3 pupils through a Technology programme that will allow them to reach modest Attainment Levels at age 12 and then abandon the subject for more traditional academic pursuits

Yet a further hazard arises in the teachers' choice of material. Virtually every school-book publisher is developing 'blockbuster' sets of books which embody a Design and Technology programme for one or more of the key stages. Whilst there are notable exceptions, the modest ambitions of some of these enterprises are worrying. Yet they are designed to enable teachers to 'cover' the four attainment targets in Design and Technology.

Recently the author, reviewing one publisher's programme in the *Times Educational Supplement* (22.11.91) commented

The style of the booklets is unashamedly pop culture, the design of trainers is a major theme and current Coke adverts are analysed for their 'appeal' (without any attempt to consider the nutritional content of the product). We are then into fashion clothing, the music charts, cinema and television programmes and even activities such as 'How to design a happening'.

More worrying is the undemanding nature of most of the activities; reasonably able children could zip through most of the booklets in a week. One has to ask: 'Is this all there is in national curriculum design and technology?' Most of the content of many of the courses is devoted to teaching how to become a more sophisticated consumer – and the majority of children manage to achieve this quite well without the aid of the national curriculum. It is all so simplistic and obvious that a trained teacher is hardly needed to deliver it.

Alas, the doubts are not only based on the publisher's efforts and the practising teachers who generate them. The HMI report on technology at Kingshurst City Technology College, with one of the best-resourced technology faculties in the country, is depressing. It observed:
'There is confused terminology in College documentation and significant areas of weakness in the teaching of designing and making. Lessons are not built around a clear rationale or structure; work in craft, design and technology is little more than a series of projects and offers little progression in skills or concepts.'

Yet another concern linked with these is the difficulty many primary school teachers are experiencing in delivering Design and Technology at Key Stages 1 and 2. We have already referred to this in Chapter 3. The reports by the Exeter University team (Wragg *et al.*, 1989; Bennett *et al.*, 1992) went so far as to label Technology as the one National Curriculum subject over which primary school teachers faced the greatest difficulty, a difficulty virtually constant throughout the period between the two surveys reported. There is a significant risk that Government, concerned with the allegedly low standards in the core subjects (English, Maths and Science) in primary schools, may be tempted, in view of the difficulties being experienced in delivering primary school technology, to reduce or even abandon work in Technology at Key Stages 1 and 2. The abandonment of Standard Attainment Tasks at Key Stages 1 and 2 in Technology announced in 1991 caused much concern. (Though at the time of writing Teacher Assessment at Key Stages 1 and 2 remains in place.)

The report for the Engineering Council by Smithers and Robinson (1992) calling for more coherent, practically based school technology is likely to cause further uncertainty.

Yet despite these clouds in the sky, the tone of this book remains cautiously optimistic. Technology education has had a glorious ten years. From its modest roots in CDT, Home Economics, Business Studies and Technical Drawing, it has been transmogrified into an acclaimed, coherent new subject, mandatory and with a 10 per cent stake in the curriculum of pupils age 5–16.

Real changes have unquestionably occurred in the status and recognition of Design and Technology and practical and vocational education. They have occurred in England and Wales and also in the United States and Canada, in Australia and in much of continental Europe.

To continue this movement to achieve full integration and parity of esteem by pupils, parents, employers and teachers themselves is the purpose of the book. To realise it all who read it must help the move forward. The success, of you, the readers, will be the author's true reward!

References

Bennett, N., Wragg, E. C., Carré, C. G. and Carter, R. (1992) 'A longitudinal study of Primary School Teachers' Competence in and Concern About National Curriculum'. *Research Papers in Education*, Vol. 7, No. 1.

Breckon, A. M. (1990) 'Design and Technology in the National Curriculum'. *Design and Technology Teaching*, Vol. 22, No. 1, pp. 4–11.

Eggleston, J. (1991) 'A Fading Flower'. *Times Educational Supplement*, 22 November, p. 16.

H.M. Government (1991) *Education and Training for the 21st Century* (2 volumes). London: HMSO.

The Independent (1991) 'Schools Charter'. London: The Independent Newspaper.

Layton, D. (1991) 'Key issues – the way forward', lecture, Warwick University Conference on Assessing Design and Technology, unpublished.

National Council for Vocational Qualifications (1991). Letter from T. Eggar (DES) and R. Jackson (Minister of Employment), unpublished.

Roberts, P. (1990) 'A subject and its objects'. *Times Educational Supplement*, 19 October, p. 26.

Smithers, A. and Robinson, P. (1992) *Technology in the National Curriculum*. London: The Engineering Council.

Wragg, E. C., Bennett, N. and Carré, C. G. (1989) 'Primary Teachers and the National Curriculum'. *Research Papers in Education*, Vol. 4, No. 3, pp. 17–45.

Index